Peter and Ar.

Translated from German by Barbara Mulcahy

Scanning and Printing

Focal Press
An imprint of Butterworth-Heinemann Ltd
Linacre House, Jordan Hill, Oxford OX2 8DP

A member of the Reed Elsevier plc group

OXFORD LONDON BOSTON
MUNICH NEW DELHI SINGAPORE SYDNEY
TOKYO TORONTO WELLINGTON

First published in Germany by Addison-Wesley GmbH 1991
First published in Great Britain 1992
Reprinted 1995 (twice)

British Library Cataloguing in Publication Data
A catalogue record for this book is available from the British Library

ISBN 0 240 51400 9

Printed and bound in Great Britain by
The Bath Press, Avon

Contents

Preface 9

Chapter 1 **A few frank words...** 11
This book was the result of annoyance! 11

Chapter 2 **Desktop publishing: a revolution for everyone!** 21
What is DTP? 21
An unusual application 22
How this book was created 24
Printing – up to now 25
Printing today – with DTP 27
Comparison of the conventional system with DTP 30

Chapter 3 **The Scanner** 33
General 34
Output resolution in greyscale pictures 36
Memory capacity required for greyscale pictures 37
Optical resolution of the scanner 37
Grey-level range or density range of the scanner 38
Output resolution for line drawings 39
Lenses, mirrors and sources of light 40
Lines and distortion in a picture 42
Scanner software 43
Types of scanners 44
Summary 45

Chapter 4 **Obtaining text and archiving** **47**

ABC

Text recognition (OCR) 47
Archiving of documents 49

Chapter 5 **Scanning drawings** **51**

General 51
Sanning resolution for drawings 51
Image size and scanning resolution 53
Raster/vector conversion (auto-tracing) 54

Chapter 6 **Scanning photographs** **57**

Grey levels cannot be printed! 59
What is the structure of a printed picture? 60
What is a raster? 61
Conversion of centimetres into inches 61
It depends on the light: the printer cannot do it! 64
What are laser standard printers and scanners capable of? 70
How many grey levels can be printed? 76
Rasterising/dithering 76
Scanner data format: TIFF must be used 78
Setting the raster 79
The raster counter 81
Difference between scanning and image resolution 81
Optimisation factor for the calculation of image resolution 82
Calculation of the image or scanning resolution 82
Memory requirement 85

Chapter 7 **Scanning rasterised pictures** **85**

How are moiré patterns created? 85
Scanning rasterised originals 88
Reduction of data with interpolation 89
Data reduction and smoothing 90
Mean value formation 93

Chapter 8	**Camera-ready copy and photosetting**	**95**
	Which camera-ready copy?	95
	Camera-ready copy from your own printer	96
	How is professional camera-ready copy produced?	97
	Photosetting	101
	Specifications for the photosetting studio	103
	What the photosetting studio needs to take into account	104
	How does a photosetter work?	105
	What is PostScript?	110
Chapter 9	**Printing**	**113**
	A new encounter	113
	Quality with DTP	314
	What is light?	116
	How do we see light and colours?	117
	Methods for the creation of colours	118
	Colour separation	118
	Colour copying with the computer	119
	The special laser printer	121
	Offset printing machines connected to computers	122
	Densitometer	123
	Which raster width does the eye prefer?	125
	Printing with offset machines	131
	From film to printing plate	133
	From tone increase to quality control	135
	The printing firm order	144
	Which printing medium for which purpose?	145
	Various printing processes	150
	Paper	151
Chapter 10	**Processing greyscale pictures**	**153**
	Important terms	155
	Correction of tone increase	156
	Scanner calibration	159
	Screen calibration	161
	Paths to success	171
	Tips for optimum-quality pictures	175

We are processing a picture 195
Miscellaneous 228

Chapter 11 Pictures from video cameras 233

Multi-media is the keyword 233
Notes on quality 234
Calculation of raster width and image size 235
Printing video images in practice 237
Printing colour pictures 240
Video pictures by means of satellite 240
Video standards 241

Chapter 12 Scanning and printing of pictures: summary and tables 243

General information on the scanning of greyscale pictures 243
Drawing and retouching 245
Checklist for greyscale scanning 246
Quality of paper 248
Recommended raster widths 249
Calculation of actual raster widths 253
Formulae for greyscale pictures 254
Conversion of inches into centimetres 254
The printing order 255
Correction of tone increase - General 257
Correction of tone increase: positive photosetting 258
Correction of tone increase: negative photosetting 260
Photosetting order 262
Grey levels as logarithms and percentages 263
Line drawings: summary 265

Chapter 13 From PC user to graphics designer 267

Basic principles of layout 267
Advertising 270
Page planning for newspapers 271
Fonts and dimensions 272
Text and serifs 275
Blocked text or left justified? 277

Letter widths and word spacing 277
Leading (line spacing) 277
Kerning 279
Widows and orphans 280
'Artistically distorted' text 280
Spaces and quads 280
Character set and ASCII table 281

Chapter 14 A DTP workstation 291

Which requirements must a DTP system meet? 291
Monitors and graphics cards for DTP 293
Monitors and graphics cards for word processing 294
The computer system 295
The printer 297
The scanner 298
DTP software 304
Ergonomic viewpoints on the workstation 305

Index 311

Preface

Dear reader,

First of all we would like to thank everyone who helped in the creation of this book, but unfortunately it is not possible to mention all of them by name here. Nevertheless, we are especially grateful to:

Mike Cash, Senior Acquisition Editor at Butterworth Heinemann, who was committed to having this book published and we would like to thank him for his outstanding cooperation. He made it possible for us to realize our ideas not only in the production of the book but also in creating the design.

It was only possible to prepare the book in this form with the help of many people whose interest in the subject motivated them to give their assistance willingly. Above all, special mention must be made here on the extraordinary commitment of **Karl R Kowalczyk**, Head of the Printing Department at Heidelberger Druckmaschinen AG, who gave us advice and help, and provided considerable support through useful notes and ideas.

Futhermore we are grateful to **Barbara Mulcahy** for her encouragement in translating the book. Within a short time we had a perfect translation!

In addition, many thanks are due to the freelance copy editor **Neville Hankins** for his useful advice and hints.

Also, **Sal Chaffey**, Desk Editor at Butterworth-Heinemann for the editorial advice.

Finally we are grateful to **Ulrike** without whose emotional and editorial support this book would not have been possible.

It has been wellknown for some time that desktop publishing has revolutionised printing. However, desktop publishing (DTP) still causes considerable controversy which became apparent to us when preparing the German edition of this book. Many experts from the printing industry are still cautious about DTP.

It is understandable why experts have difficulty in accepting that DTP users are able to create camera-ready copy of professional quality at their desks, without the training and experience which took the experts years to gain. Their general opinion is 'Why are we still needed, if everyone can do everything themselves?'. This book shows how it is possible to integrate good-quality pictures relatively simply into documents.

The way in which DTP users and printing experts work together needs to be changed. The fear of experts that they may become superfluous is completely unfounded. DTP users may have many technical aids available, but that does not mean they possess the necessary creative ability. As far as the printer is concerned, professional printing will still be carried out and probably to a greater extent than previously. Therefore all that is needed is for the various groups involved to develop a new way of working together.

In addition, concerning the structure of this book, we would like to mention that we have tried as far as possible to keep individual themes separate, so that if you are only interested in one particular theme you do not have to read the entire book. Hence some specific basic relationships need to be repeated in the book. However, the repetition of certain aspects is beneficial for those of you who are dealing with this subject for the first time, as it will be easier for you to understand the relationships.

Peter and Anton Kammermeier
and the MEGASYS DTP-Team

Chapter 1
A few frank words...

This book was the result of annoyance!

Everyone is talking about desktop publishing (DTP). However, very little concrete information or help exists on the complete process (from the scanning of pictures through to photosetting and printing). Even manufacturers of DTP hardware are often unable to give sufficient help and information in books is often too general and sometimes even incorrect. For example, in a certain expensive book on DTP, it says that a good-quality picture cannot be obtained with a 300 dpi scanner and hence the limits of DTP have been reached. In addition, it says that if only 300 dpi resolution is used it is not possible to produce more than 50 lpi (lines per inch). As a result of comments like these, people strive for higher and higher resolutions. But DTP users be warned – this is incorrect!

The picture on this page was scanned with a standard DTP scanner at a resolution of 100 dpi!

(Photograph: Jutta Hoffmann)

Your scanner is capable of this: a photograph scanned at 100 dpi and still good quality.

If authors write about DTP, but have never scanned a picture correctly, photoset and printed it, then their information remains theoretical and of little use in practice. Usually such authors will summarise by describing what cannot be achieved with DTP. They will describe how to use process cameras, how pictures are enlarged and cut, and how photographs are wet processed

It is guaranteed that all photographs in this book were produced using a standard DTP scanner! This photograph shows the old opera house in Frankfurt/Main.

with developing fluid, fixer and running water. As in days gone by, the talk is all of cut and paste, with no mention of a scanner. On the other hand there are the DTP sales people who say everything is very simple. Brochures, specialist magazines and manufacturers promise you that it is really straightforward to produce perfect printing results. Well, all possibilities between these two extremes exist. We would like to know exactly what is true. But as often happens, the more you occupy yourself with a subject, the more complex it becomes. For example, you discover that scanner manufacturers are effective at selling, but do not really know the basic principles themselves. Trying to obtain help and information can be very difficult. In answer to specific problems you will get an answer, but it is often vague and sometimes completely wrong. Bad results are usually blamed on the equipment and technology

(Photograph: AUDI AG)

Nowadays with laser printers even 2000 dpi-quality pictures can be obtained.

not being sufficiently advanced. In reading this book, you will discover that it is not the technology which is to blame for poor results, but the 'know-how'. Even with a cheap handheld scanner it is possible to obtain superb results. The subject of scanning takes up considerable space in this book, but all the other important aspects of DTP are also mentioned. It would not be worth while to describe these aspects only partially and thus omit to clarify the relationships between them.

This book describes the basic principles involved in DTP and provides concrete notes and tips. A large part of the book is devoted to the subject of 'correct scanning and printing'. In order to obtain optimum results when scanning and printing photographs, it is important to have a basic understanding of the technology.

However, we will describe not only the processing of pictures (even those which have been taken by video camera), photosetting and printing, but also the whole spectrum of ergonomics, typography, advertising and the purchase of a DTP system. Some of these issues will be looked at from a different point of view: we will not examine the functions which have failed, but instead we will check whether the basic functions have been implemented correctly.

(Photograph: Jutta Hoffmann)

Anyone who is capable of taking photographs can also create perfect artwork.

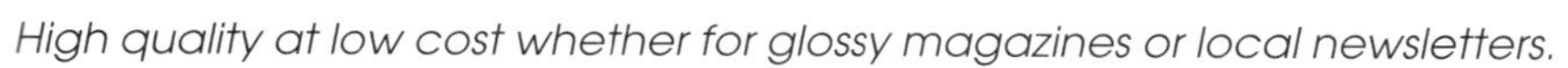

High quality at low cost whether for glossy magazines or local newsletters.

Even if a cheap handheld scanner is used, it is possible to obtain good quality.

It did not work until now

Often people do not consider calibration, dot enlargement and the proper data format. Did you know that too high a resolution is usually used for scanning and not one that is too low, or that you must not rasterise images when scanning? Did you know that a scanner with a resolution of 1000 dpi actually only has a resolution of 300 dpi? Often, all you consider is the 'resolution'. Whether the real resolution (physical or optical) or the simulated resolution (calculated) is being dealt with is usually ignored.

We had to carry out comprehensive research, experiments and tests in order to provide information on DTP which is not superficial, but instead a real help. As a result, this book will provide you with concrete tips and exact values and formulae, which should help you to produce photographs of the highest quality on your DTP system. After reading this book, you are unlikely to make the mistake of scanning a photograph at 300 dpi if you are going to print it on a laser printer. Furthermore, information for professional applications (photosetting at 2540 dpi) is included. Recently photosetters have become cheaper to the extent that even a private individual can afford one.

(Photograph: ETAP Information Technology, Belgium)

Grey levels are important whether on double-page or standard VGA monitors.

Using a concrete example, we will describe the advantages of DTP for the user and the problems which exist for printers who ignore this new development in publishing. In order to evaluate the capabilities of DTP properly, we will describe the traditional methods and compare them with the new technology. Incidentally, this book, and all its photographs were created completely on a DTP system.

Even scanner manufacturers should read this book

Our examination is primarily aimed at the DTP user. However, DTP component manufacturers are also addressed with advice on how their products might be improved in one form or another. In preparing this book we not only researched

(Photograph: Stefan Dintelmann)

In the future people will say 'Poor picture quality! Why don't you use DTP?'

(Photograph: Jutta Hoffmann)

It is not the price of the hardware that matters, but knowledge. Incidentally, the photographs on these two pages were printed using elliptical points.

present knowledge, but also designed new solutions in problem areas. For example, we describe a process here for the first time which we developed to avoid the undesirable moiré patterns (interference patterns) that occur when scanning pictures which have already been rasterised, for example pictures found in magazines, brochures, etc.

Diverse groups work together to consider the subject of image processing, though often they cannot completely agree with each other. For example, the DTP user does not understand enough about half-tone increases in printing (pictures are made darker). The printing expert understands this and has considerable experience in how to handle his or her machine, but is unable to help the DTP user enter the correct values into a computer.

Since a lithographer normally creates the camera-ready copy, the printer sends the DTP user to a lithographer. Here the same scenario occurs – problems arise with the terminology, for example when density, lighting, shade and tone increase are mentioned. If a term is mentioned which both the DTP user and the lithographer understand, it may mean something different to both. Therefore, in the end, the lithographer also cannot provide any values to help the DTP user attain successful results.

This is the big dilemma. The training and experience required for the conventional method is not easily converted into values for a computer. This is because the printing expert and the lithographer do not have the necessary computer expertise and do not understand the technical relationships involved in a DTP system. The DTP user has the advantage that even without substantial training, provided the correct values are entered, he or she can use the computer to make a good job of the task.

This 'not understanding' accompanies the DTP user at all levels. A DTP sales person rarely understands anything about the 'black art'. Therefore the general opinion always reached when pictures have been scanned too dark, too flat or too coarsely is that DTP is not yet sufficiently advanced. This is not due to DTP technology but to the incorrect use of devices, and this will be demonstrated in this book.

Even manufacturers of DTP components concentrate mainly on their own affairs and do not notice when features do not function optimally.

Chapter 2
Desktop publishing: a revolution for everyone!

As was once the case with filmsetting, DTP is proliferating. However, many people have not yet taken much notice of it. 'Desktop publishing', also known as 'computer publishing' or 'electronic publishing', is a revolution for everyone and not just the printing industry.

What is so revolutionary about DTP? Is it the new things that printers and graphic designers are able to do using it? No! What is new is that anyone can afford a 'printing firm' for their own desk and that everything is very simple to use. Many issues which were addressed either not at all or only insufficiently, because they were too expensive or required too much effort, are no longer a problem. Furthermore, the way in which we work with graphic designers and printers has changed.

What is DTP?

In simple terms, it is no longer necessary to cut and paste at a light table. Instead anyone can create their own camera-ready copy on a personal computer. By just pressing a button, high-quality finished documents can be produced from a printer.

DTP only describes the function of arranging texts and pictures on a screen. Texts and pictures are shown on the screen at a size of almost 1:1 just as they will be printed later. Text may be created using any word processing program. Pictures (photographs, line drawings) are reproduced using a scanner. However, pictures may also be obtained by means of a video camera or produced using special programs, such as presentation, calculation or drawing programs. Photographs of professional quality may be produced using a raster of 150 lpi. Using the new generation of photosetters they may even be produced with a raster of 200 lpi (80 lines per centimetre (L/cm)).

If you desire perfect quality, you no longer send the printing firm text and pictures on paper, but instead you send the information by means of a data storage medium. The printer is no longer responsible for the creation and setting of work, but now only reproduces it. The advantage for the customer is that typing errors do not creep in during the printing process. It is cheaper and simpler. Even high-quality pictures may be entered into a computer and processed.

An unusual application

To enable you to understand the 'unlimited' capabilities of DTP a little better, we will describe an unusual application. By pressing a button a catalogue with more than 800 pages was created literally within a few minutes. Text, drawings and photographs were integrated automatically.

The creation of this catalogue used to be laborious and required considerable effort and expense. Now it can be created just by pressing a button and without any manual work being carried out. It used to be necessary to plan ahead as the preparation required 3 months, but now the catalogue may be created at the last minute.

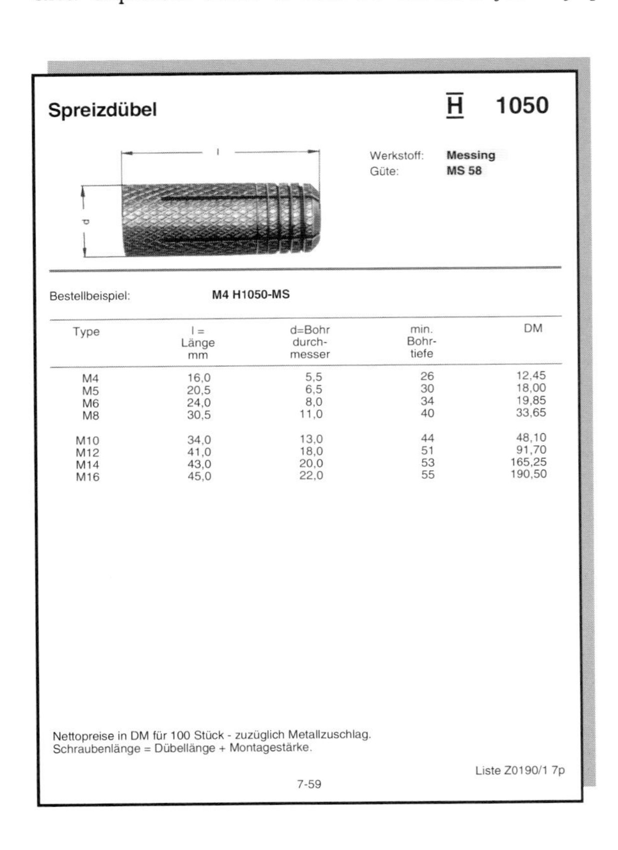

Spreizdübel **H 1050**

Werkstoff: **Messing**
Güte: **MS 58**

Bestellbeispiel: **M4 H1050-MS**

Type	l = Länge mm	d=Bohr durch-messer	min. Bohr-tiefe	DM
M4	16,0	5,5	26	12,45
M5	20,5	6,5	30	18,00
M6	24,0	8,0	34	19,85
M8	30,5	11,0	40	33,65
M10	34,0	13,0	44	48,10
M12	41,0	18,0	51	91,70
M14	43,0	20,0	53	165,25
M16	45,0	22,0	55	190,50

Nettopreise in DM für 100 Stück - zuzüglich Metallzuschlag.
Schraubenlänge = Dübellänge + Montagestärke.

Liste Z0190/1 7p

7-59

Many people were of the opinion that it should be left well alone: 'It won't work with DTP' was the general feeling. However, a few did not give up so easily. Their courage was justified since within a few weeks a functional concept had been produced.

It is possible with Desktop Publishing!

In this application, at regular intervals a comprehensive catalogue/price list with more than 800 pages was published. The

conventional process via a printing firm was very time consuming and required considerable effort. Whether it was a picture that was placed incorrectly or the infamous last-minute changes, the time and effort required to create an error-free catalogue was very expensive. It made more sense to jump on to the 'desktop publishing' bandwagon and create the complete catalogue using a personal computer. This was especially so since all the prices had already been calculated using a computer and hence were already available on a data storage medium.

At the press of a button...

The task now existed of creating the catalogue without manual intervention simply by 'pressing a button'. Only in this way could the camera-ready copy for an error-free and comprehensive catalogue be produced within a few days.

Therefore, existing data needed to be prepared and created using DTP software. But, as often happens, it is the little things that cause problems. For example, one problem was that the layout (text and pictures) for the left and right hand pages was different. If one new page were to be inserted, all the following pages would need to be altered. It is of course obvious that this cannot be done manually for a catalogue of several hundred pages.

The problem was solved by using a control program which, accessed from a user menu, combined all the required steps. The most important function of this program was the step 'Organise complete catalogue'. By pressing a key on the personal computer, all the text was organised, files for the DTP system (catalogue pages) were automatically created and the pictures and text were inserted accordingly. The formatting for the left and right hand pages was carried out at the same time.

The pictures required for the catalogue were stored before they were needed in a so-called picture database in the personal computer. They had been obtained either by means of a scanner or from a drawing program. The correct pictures were then automatically placed at the correct positions in the catalogue. Those pages on which items were to be placed could be determined very easily. A list was produced, also on the computer, where it was noted which product should be on which page with which picture.

... and within a few minutes 800 catalogue pages!

It thus took only minutes to produce a catalogue with over 800 pages on the personal computer. A sample was then printed using a laser printer and the information sent to the printing firm via a data storage medium. Next the printing firm photoset this data on to film and the complete camera-ready copy (films) was produced. Text and pictures were of course completely arranged on this camera-ready copy. It took only

a few days from the initial work through the last change in price to the completed camera-ready copy. By using this process, it was primarily ensured that errors did not creep in and that the document did not require manual creation.

Problems exist with hardware too!

No one considered that problems would arise in using standard hardware and software. For example, the selection of a high-performance scanner caused problems, even though, according to the brochures, it should have been a really simple matter. It was the scanner software which actually proved to be the real problem.

For DTP it is important to have a powerful computer system. Graphical applications should not require patience. Above all, the performance and smooth operation between individual system components is extremely important. A 386 computer was used with a 330 Mbyte hard disk; 40 Mbyte SyQuest cartridges were used for backing up the data and transferring it to the photosetter. As in all components, but especially in the case of the monitor (the connection between man and machine), care was taken to have an ergonomically perfect model. An anti-dazzle, flicker-free ETAP double-page monitor with 75 Hz image scanning frequency was used. This system is now used not only to create catalogues, but also more and more for a diverse range of brochures and other documents.

How this book was created

Next, we will briefly describe how this book was produced using a DTP system. First the text was typed into a personal computer. We used the word processing program WORD from Microsoft. WORD has many advantages, especially for the creation of longer texts: the more text created, the more important the classification function of WORD became. At any time the headings alone could be viewed and hence an overall perspective was always maintained. There was also no problem in moving chapters, since moving a heading moved the corresponding text with it. When most of the text had been created, it was reproduced on paper by means of a PostScript laser printer to gain a better overview, to monitor the text and to carry out corrections.

The layout was then created using the Ventura Publisher DTP package. First the basic layout such as the book size and page margins was set. Naturally it was worth while trying a few different ideas and printing a few samples. Finally the text typed using WORD was loaded into Ventura. For reasons of simplification and speed, the text was divided into separate pieces for each chapter. During the loading of text

Ventura automatically created all the required pages with the set format. As several formatting features had already been included when creating the text (e.g. for headings), the basic structure could be created within a few minutes.

Next, creativity and hard work were required. The fonts, text sizes, headings and subheadings to be used needed to be set. Here, for example, a parameter in a menu was simply changed and all the headings in the book acquired a new format. The result could immediately be seen on the screen. Initially several samples of a few pages were printed in order to evaluate the final result. Once all the details had been decided upon, the actual creation could be carried out.

The pictures for individual chapters were entered into the computer by means of a scanner. As the final sizes of pictures were not to be determined until later, all pictures were provisionally scanned at a very low resolution. They were then positioned and printed by a laser printer so that they could be checked. Next we worked from page to page and entered the pictures at the correct size. Text and pictures were shown at a size of 1:1 on the screen. In addition, pieces of text were changed, deleted or inserted. Finally, it was time for a complete sample print. For this purpose a complete publication was created, i.e. all the individual chapters were put together. Before a well-earned break, the table of contents was created automatically. Left completely alone, the DTP program now had to print the complete book using a PostScript laser printer.

By the next day we had recovered – fortunately perhaps, as we did not really like several aspects of the book. But thanks to DTP it did not really matter. Changes could be carried out relatively simply. All the pictures used were scanned again. As the sizes for all the pictures were set, it was now possible to scan using accurately calculated resolutions.

Once everyone concerned had given a general OK to the book, the glossary was created. Then came the big moment: instead of sending the data to the printer, the data was copied on to 40 Mbyte magnetic tape cartridges. These cartridges were sent to the photosetter. Within a few days the film came back. After a quick check on the quality of the pictures the film was sent to the printing firm.

Printing – up to now

Everything is done the wrong way round, i.e. with the pages back to front. Even in modern filmsetting, work is carried out at a light table with the text and pictures being cut and pasted with the pages back to front (mirror image). What are the individual steps in this classical process?

Text creation by the author

First, as always, the text is created. Often the text is still presented on paper by the author.

Layout planning by the designer

The layout designer then plans the format for the publication. The designer sets the so-called type area, determines the space required for pictures and text and marks the text with typesetting instructions. The typesetter is provided with information regarding which font and size of text should be used for typesetting.

The typesetting studio recreates the text

The processed text is sent to the typesetting studio. Here the typesetter reproduces the complete text and at the same time provides control commands for the typesetting system. So-called galley proofs are created. Galley proofs are phototypeset lines of text on paper.

The galley proofs are corrected by the author

The galley proofs are read and corrected either by the author or by a copy editor and errors which the typesetter missed are marked.

The layout designer pastes up the proofs

The layout designer pastes up the galley proofs and determines whether the text needs to be shortened or lengthened.

The author has to shorten texts which are too long

Such changes to the text are once again carried out by the author. After the author has confirmed that the text is correct the finished layout is returned once again to the typesetting studio. The typesetter carries out the specified changes to the text.

Imposition planning

The imposition planner converts the sketched layout into an exact reproduction copy. This includes cutting marks for the format size, text proofs (made up correctly), colour specifications, status and section specifications for pictures.

Final control by author

Before everything is finally sent to the printer, the author once again determines whether all the material is correct and, if not, provides corrections.

Any other requests?

Now just imagine what would happen if the author wanted further changes to be made and the whole process had to be started from the beginning again...

The litho is created

The lithographers now require all the material: page layout, pictures, slides, etc., in order to carry out yet another stage on the way to the end product. So that the individual parts are available on film, the page layout is photographed using a process camera and pictures and slides are scanned or, using a process camera, rasterised. These individual parts, pictures and photographed page layouts are assembled manually in accordance with the layout specifications. This results in the final litho: page reversed positive films.

Printing preparation

At the printing firm the various processes merge again. Whether the processes were carried out conventionally or by using DTP, the printing firm then produces printing plates from the film and printing may commence.

Printing today – with DTP

Everything is so simple with DTP: number of columns, size of margins and line spacing are easily changed and the result is immediately visible on the screen. Another button is then pressed and the work is complete. Whereas before you had to busy yourself with scissors, scalpel and glue, now all you have to do is place texts and pictures on the screen using a mouse. The design may be changed repeatedly until it is acceptable. Imposition planning is no longer needed, as each design is an original and the level of quality may be controlled. With DTP both the time-consuming and decision-intensive steps are concentrated within your own firm or department. Text is created once only and so possible sources of introducing new errors are removed. The layout on the screen provides immediate information as to whether a text needs to be shortened or lengthened. You become more flexible and the infamous last-minute changes are no longer a problem.

Using DTP you can maintain complete control over the entire production. As manufacturer of the camera-ready copy you are in charge of the complete process of typesetting, producing graphics and carrying out the page layout. With this complete control you become extremely flexible and able to react quickly to changes. Savings in time and cost are enormous. All the processes involved in creating camera-ready copy may be carried out at your own desk. There is no time-consuming postage to and fro, no need to proof-read repeatedly and no need to check the layout several times. Many tasks which had to be carried out elsewhere are either not required at all or may be carried out by yourself. Taking all these advantages into account, the cost of investment is relatively low. In addition, if you consider, for example, that the

result obtained for smaller publications is considerably improved, or that an advertisement for a daily newspaper may be completed 5 minutes before the deadline, then it is difficult to imagine working without DTP.

DTP – simple?

Yes and no. The technical capabilities of DTP lead people to believe that nothing could be easier. However, DTP is not as simple as the advertisements in colourful brochures would like us to believe. Let us examine the issues from a critical point of view for a moment. DTP is a very useful computer application, but also one which leaves you with the responsibility for what you print. You alone produce the layout and make the decisions. DTP often combines the activities of author, publisher, copy editor, editor, typographer, graphic designer and computer expert into one.

Especially when you begin using DTP do not overestimate your abilities. It is not just a case of learning how to produce page layouts. Though DTP is simple and perfect to use, it still needs to be learned. A computer novice without expert support will experience more frustration than joy. However, an experienced computer operator should also ensure that he or she receives sound advice. It is often minor revelations or even just a sensible introduction to it that will make DTP a successful experience.

What does a DTP process look like?

Creation of text

First the text is required. For example, for a leaflet advertising special offers you would need product descriptions and prices. If these product descriptions and prices are already stored in your computer, you can transfer them and save having to retype the information. Otherwise you will need to type the required text as before, not by using a typewriter, but instead much more conveniently by using the computer. The text may be changed, corrected and moved without difficulty. If you are using a professional word processing program, many functions for the checking of spelling, automatic creation of the contents, etc., are available. You may then either send this text to a DTP service /printing firm for further formatting, safe in the knowledge that no more errors can creep in, or carry out the layout of the leaflet yourself.

Obtaining pictures

You may either create pictures yourself (e.g. with a drawing program, table/calculation program, CAD system, etc.) or input them into the computer by means of scanning original pictures.

Layout

Now you may begin immediately! Using a DTP program the text and pictures which have already been created may be combined and laid out directly on the computer. On the screen you can view the result immediately, exactly as it will be printed. Pictures may be moved, reduced or enlarged and the text may be rearranged, with changes to size and line spacing. These changes may be applied repeatedly until the result is acceptable and the text fits within the required number of pages.

Sample print

Once the layout has been completed a sample may be printed. Even printing experts are astounded with the quality of the printout from a PostScript laser printer. There is only one task which the laser printer cannot do, namely print high-quality photographs. It is not possible to attain more than newspaper-quality photographs with a laser printer. However, the layout may always be checked using a laser printer.

Inexpensive printing

If the printout from the laser printer meets the quality requirements sufficiently, it may be inexpensively copied or reproduced by the small offset process. In this case no film is produced, but printing is carried out instead using the paper camera-ready copy from the laser printer. The quality of photographs may be improved by requesting a copy shop or printing firm to rasterise and assemble the pictures. A true DTP user, however, will send the text and pictures to a photosetting studio in order to have them photoset.

Professional printing

Once a sample has been printed on the laser printer and the layout meets the requirements, a so-called PostScript printing file is created. This file is sent via diskettes (for example) to a photosetting studio or printing firm and then photoset on to a professional output device (photosetter). Whilst a laser printer has a resolution of only 300 dpi, more than 2540 dpi can be attained using a photosetter. As a perfect result you will receive the camera-ready copy on film or paper. The pictures are of course included in this camera-ready copy.

The advantage

The advantage of producing documents using a DTP system is that from the initial creation of text right through to the complete layout, everything can be achieved by yourself. Errors due to the incorrect typing of text or incorrect arrangement of pictures, for example, will be avoided. Subsequent changes are simple with a DTP system, as all necessary information is stored within a data storage medium and only the required changes need to be applied. Once the changes have been made photosetting will need to be carried out again and the new camera-ready copy will then be complete.

Comparison of the conventional system with DTP

Creation of final Print			
	Conventional printing		DTP system
1.	Text is handwritten or typed using a typewriter or word processor.	1.	Text is typed into a personal computer.
2.	Layout planning. Changes are time consuming and expensive.	2.	Layout planning. Changes can be made at any time.
3.	Text is marked on the paper for the typesetter (labels, font, text size, etc.).		N/A
4.	The text is sent with setting details to the typesetter. The text is copied taking into account the setting parameters (typesetting) and then photoset (galley proofs). It is then proof-read and if necessary photoset again.		N/A
5.	The author receives the galley proofs to check.		N/A
6.	The printer receives the corrected galley proofs, carries out the corrections and photosets again.		N/A
7.	The author then receives the galley proofs again to recheck them.		N/A
8.	Pictures and drawings are selected.	3.	Pictures and drawings are selected and are then either entered by means of a scanner or created by means of a computer program.
9.	Pictures and drawings are reproduced by a lithographer.		N/A
10.	Galley proofs and reproductions (pictures) must be cut to the right size for each page and then individually mounted (pasted) on to each page.	4.	Inserting page breaks and arranging pictures are carried out on the screen. Changes may be undertaken immediately.

11. Owing to changes which occur during the paste-up (setting of pages) photo-setting must be carried out once more at the printers.	N/A
12. The new galley proofs are cut and pasted again.	N/A
13. The paste-up is finished and the final check is undertaken.	5. The finished pages are printed by means of a laser printer so that they can be checked.
14. If corrections are required the steps above are repeated.	6. Corrections can be undertaken immediately.
N/A	7. All the printing files are sent to a DTP service or to a printing firm to be photoset.
15. The finished camera-ready copy is sent to the printers.	8. The finished camera-ready copy is sent to the printing firm.

Advantages and disadvantages

Conventional systen	DTP system
Epensive systems	Comparatively cheap systems
Slow process	Quick, as everything included within one system
Use of external organisations: typesetting, creation of graphics and printing are often carried out by different organisations.	Complete responsibility within own firm; integration of text creation, graphics and camera-ready copy within one process
Little control over the complete process	Complete control over each step
Your contribution to the work is only small.	You need to contribute the most work.

Chapter 3

The Scanner

By using a scanner, pictures (photographs and drawings) may be entered into the computer. A large part of this book deals with what needs to be considered if you wish to attain good-quality pictures. As far as scanning and image processing are concerned, 'know-how' is the most important. There are many factors which need to be considered. What makes matters more difficult is that until now many of the basic relationships were not very well understood and only scant fundamental information existed.

The various components of the process, from the scanner to the program and printer, must be compatible with one another. Considering only a single component, such as the scanner, will not lead to exceptionally good results. However, the most significant errors may have already been produced when scanning. This chapter hopes to make the scanner a little more familiar. Important details and practical advice on image processing which need to be taken into account will be addressed in more detail in later chapters of this book.

Drawings and photographs are entered into a computer using a scanner.
(Photograph: Hewlett Packard)

Please note that all the pictures in this book were obtained using a standard DTP flat-bed scanner!

Basically, a scanner operates as follows. Light from a suitable source is shone on to the original document being scanned. The light reflected from the document (e.g. photograph) is directed via a mirror and lens system to light-sensitive sensors, so-called CCD sensors (charge-coupled devices). Each sensor may be considered as a type of tiny camera. These sensors detect the brightness (grey level) of the reflected light and convert it into a voltage.

This voltage is digitised by means of an analogue/digital convertor (A/D convertor) and the data produced is then passed on to the computer. As a result the computer receives a value for each point on the illuminated image. The level of this value depends on the type of scanning process used and the power of the scanner.

General

By means of the scanner software it is possible to select whether

- drawings
- rasterised greyscale pictures
- non-rasterised greyscale pictures

are to be scanned and then saved.

If by means of the scanner software you opt to scan a drawing only, then grey levels will not be read. Instead only two states (0 or 1, black or white) will be detected. Intermediate states do not exist. The level (limit) of brightness of each picture dot is considered to be 'black' or 'white' and that is all that may be specified. Therefore 1 'bit' (0 or 1) provides sufficient information for each point scanned.

If you opt to scan grey levels (e.g. a photograph), then it is not merely a matter of detecting two states (black or white) as in drawings, but of obtaining a specific level of brightness for each picture dot. So-called greyscale scanners store this grey-level information directly and without need of further processing.

Once again imagine a tiny camera (CCD sensor) which travels slowly from the top to the bottom of a photograph continually taking pictures. Each one of these tiny pictures represents a grey level. If a 256 greyscale scanner is used, this grey level is stored as a number between 0 and 255 inclusive. If the CCD sensor detects a white area then the number 0 is sent to the computer; for a black area the number is 255

and for a medium grey area the number is 128. This means that, for each grey level, the scanner is able to recognise 256 different states (levels of brightness) and send these to the computer. Hence the computer must be able to store an 8 bit character (256 states) for each grey level. In contrast, as already mentioned, for line drawings only two states (black or white) exist, i.e. only a 1 bit character needs to be stored.

Depending on the type of scanner, the number of grey levels which may be detected and saved for each dot viewed are 256 levels of brightness (8 bit scanner), 64 levels of brightness (6 bit scanner) or 16 levels of brightness (4 bit scanner). As previously stated, greyscale pictures may be scanned by one of two methods: rasterised and non-rasterised. If by means of the scanner software you opt for the image to be non-rasterised (i.e. not dithered), the grey levels are stored as described above without further processing, i.e. for each grey level a number from 0 to 255 is assigned. However, scanners are also capable of rasterising (or dithering). Rasterising or dithering means that the grey-level information from a photograph is processed further, in order for it to be printed. The issue that standard printers are not able to show grey levels, simulated by dots of varying sizes, will be discussed in more detail later in this book. As a printer cannot print dots of varying sizes, grey-level dots of varying sizes are converted into several smaller dots of the same size.

As you will discover, you do not require the rasterising facility. Hence the fine distinction between rasterising and dithering is of no importance here. We will only say this on the subject: with rasterising different grey levels are simulated by dots of varying sizes (change in amplitude); with dithering all the rasterised dots are of the same small size and the different grey levels are attained by means of varying the offset between the dots (change in frequency). Many variations of rasterising and dithering exist. For example, the dot size may be changed or the dot size and dot offset may be changed at the same time.

When rasterising, i.e. when obtaining pixel images from greyscale information, all the image characteristics such as size and arrangement of the raster dots are fixed. The rasterised image cannot be changed, otherwise the arrangement of pixels changes which definitely results in the quality being reduced (e.g. moiré interference patterns). When rasterising, the final size and quality of the image (output from photosetter or laser printer) must be fixed and not be allowed to change thereafter.

Rasterising is therefore the last link in the image processing chain, particularly as the creation of a raster is mainly dependent on the resolution of the output device. Only when the image has been placed at the correct size in the DTP software can rasterising take place. The resolution of the output device is then taken into account in the DTP program or in PostScript in the laser printer or photosetter and not by means of the scanner software. Under no circumstances should images be rasterised during the scanning process.

Output resolution in greyscale pictures

When rasterising takes place the information representing the individual grey levels is translated into black dots of varying sizes. This means that the brightness levels (0 to 255) scanned are each converted into multiple black and white values (0 or 1) and then saved. This conversion is particularly carried out with such output devices as laser printers. Corresponding to the resolution of the printer a sufficient number of black pixels must be available so that each desired grey level may be created.

Hence, when rasterising it is not the scanning resolution which is important, but the resolution of the printer. If, for example, 'rasterising' or 'dithering' is actually selected via the scanner software and a scanning resolution of 300 dpi is selected, then in reality images will only be scanned at a resolution of about 65 dpi and from this an output resolution (number of pixels for the printer) of 300 dpi will be produced. The reason for this will be described in more detail later in this book.

So if you rasterise at the same time as scanning, it is the 'output resolution' you are considering as you are not scanning at the resolution set. Instead the image is being output at the resolution set. However, scanners do exist where you do not have to choose if an image is to be rasterised or dithered until it is to be saved. In this case, if a scanner resolution of 300 dpi is selected, scanning will actually take place at a resolution of 300 dpi, but on saving all the information at resolutions higher than about 65 dpi are discarded.

In order to output a photograph to a photosetter (photosetter resolution of 2540 dpi) it is usually sufficient to use a scanner resolution of about 150 dpi. The reasons for this will be described in more detail later. At this point, we will say only that in order to print the scanned 256 grey levels, the printer (photosetter) requires a resolution which is a factor of 16 times higher than the resolution of the scanner. This is because the printer has to use several pixels for each grey level. For example, when rasterising 100 scanned grey levels (real scanner resolution of 100 dpi) 1600 pixels (output scanner resolution of 1600 dpi) need to be used if all 256 grey levels are to be output. If only 25 grey levels are to be printed, then a 500 dpi output resolution may be used for 100 dpi grey-level information. This high output resolution is often incorrectly specified simply as the scanner resolution, even though scanning does not take place at this resolution. Hence, if you rasterise at the same time when scanning photographs, it is not the scanner resolution which is specified, but the relevant printer or photosetter resolution.

 *As there is no need to convert real grey levels into a pixel dot matrix, the output resolutions specified in technical data sheets from manufacturers are of little use. Rasterising must only take place by means of the DTP program or by means of the printer or photosetter, but under no circumstances during scanning.*

Memory capacity required for greyscale pictures

The widespread opinion that rasterised images require much less memory is another statement that we cannot ignore. It is not possible to compare the amount of memory required for scanning with rasterising and scanning without rasterising. The results produced are completely different and cannot be compared. A picture scanned and rasterised at 300 dpi results in an image which is only just acceptable for printing on a laser printer. An image scanned but not rasterised at 300 dpi results in a high-quality picture when output from a photosetter.

The explanation is simple. If you simultaneously scan and rasterise an image using a standard scanner at 300 dpi, this does not mean that you are scanning at 300 dpi, but that a dot raster of 300 dpi is created with respect to the output device. In reality you are only scanning at about 65 dpi. Hence you obtain nearly identical results if you scan and rasterise an image at 300 dpi or if you scan and do not rasterise at 65 dpi. Thus the argument about the amount of memory required is completely irrelevant.

If real grey levels are stored on scanning, a much lower scanning resolution may be set and yet far better picture quality is attained than with scanning and rasterising at a high resolution. Therefore there is no great difference in the amount of memory required for pictures which are stored as grey levels and those which are stored rasterised.

Optical resolution of the scanner

People are always talking about resolution, but they do not usually mention whether they mean the output resolution (calculated resolution) or the actual resolution at which scanning takes place. As a result of its hardware characteristics, the resolution at which a scanner can actually scan an original document is known as the 'physical' or 'optical' resolution. Standard DTP flat-bed scanners have a physical resolution of

300 or 400 dpi. Only a few and very expensive scanners have resolutions of more than 400 dpi. The number of sensors (CCD sensors) corresponds to the physical or optical resolution in the horizontal direction (x axis). For example, if a scanner has an optical resolution of 300 dpi (dots per inch = pixels per inch) on the x axis, this means that it has 300 CCD sensors per inch. During scanning, a slide moves across the original document and allows the reflected light to reach the CCD sensors line by line. Hence the resolution of the motor, which moves the scanning unit, corresponds to the resolution on the y axis. Scanners exist whose CCD sensors have an optical resolution of 300 dpi, but whose stepping motor which moves the CCD sensors has a resolution of 600 dpi. Some of these scanners are incorrectly termed true 600 dpi scanners, even though their optics only have a resolution of 300 dpi. In addition, it is possible to calculate the optical resolution from the number of CCD sensors. The physical resolution on the x axis is calculated by the number of CCD sensors divided by the width of the scanning unit. For example, 2250 CCD sensors divided by a width of 8.5 inches (21.5 cm) produces a resolution of about 300 dpi.

Whenever resolution is considered, it is implied that it is defined with respect to length, i.e. pixels per centimetre or inch. The main term resolution, however, also includes how accurately grey levels may be resolved. This is specified by the number of grey levels. Here it is important to know in addition whether the scanner can physically measure 256 grey levels or whether it actually measures only 64 grey levels and then extrapolates them to 256.

Grey-level range or density range of the scanner

The grey-level or density range is one of the most important characteristics of a scanner. By density range, experts mean how easily a scanner can recognise dark areas on a picture. The density range is dependent on the grey-level range and is given by the number of detectable grey levels: 256 grey levels are available with standard 8 bit scanners.

Let us recall how a scanner works. The reflected beams of light from a photograph are converted by CCD sensors into analogue voltages. These analogue voltages are converted by an analogue/digital convertor (A/D convertor) into digital characters (e.g. 8 bit). How accurate this conversion of the voltage from the CCD sensors is depends on the accuracy of the A/D convertor. Therefore 10 bit scanners which have a 10 bit A/D convertor have a larger greyscale range (1024 grey levels) than 8 bit scanners (256 grey levels). The problem currently is that all image processing programs, including PostScript, can only process 8 bits. In addition, up to 256 grey levels only may be photoset and printed. Therefore a 10 bit scanner does not seem to have any advantage over an 8 bit scanner. This is also true when the scanner software does not use the additional capabilities of the scanner.

Several times in this book mention will be made of the fact that, when photosetting and printing, pictures become darker and need to be made lighter (approximately logarithmically) beforehand. By making a picture lighter, all the light grey levels are removed and the dark grey levels are expanded. Consequently, an 8 bit scanner provides too wide a grey-level range for light areas and too narrow a grey-level range for dark areas.

Some scanners provide a remedy to this in that they have a logarithmic amplifier built-in on the analogue side, thereby making the picture lighter before digitising. In this case the digital values gained are already well distributed and hardly any more grey levels disappear due to subsequent picture manipulation. The disadvantage of scanners which have built-in logarithmic amplifiers is that you are restrained by the rigid grey-level distribution of the amplifier and cannot carry out individual adjustments. The best scanners are those which recognise 10 bits (1024 grey levels) or 12 bits (4098 grey levels). If these scanners are used, an individual gradation curve may be set at the computer and then sent to the scanner. Using this gradation curve the required 256 grey levels may be gained from the 1024 or 4096 grey levels and transferred to the computer.

As a result, whether you are using an 8 bit, 10 bit or 12 bit scanner, only 256 grey levels are then processed in the computer. However, if a standard 8 bit scanner is used when a picture is made lighter then grey levels are missing from the darker areas on the picture. This means that if 10 bit or 12 bit scanners are used then dark areas on the pictures can be transferred far more effectively. In general pictures need to be made lighter for printing reasons and some may also need to be made lighter if the originals were very dark. The reasons why photographs need to be made lighter for printing will be explained in more detail later. From the pictures in this book it is clear that, even if an 8 bit scanner is used, very good-quality pictures may be obtained. As mentioned earlier, all the pictures in this book were produced using an inexpensive 8 bit scanner.

Output resolution for line drawings

The resolution inherent in a scanner due to its construction (hardware) is called the physical or optical resolution. However, scanner manufacturers do not always only provide the physical (actual) resolution, but sometimes quote the calculated resolu-

tion (output resolution) too. Sometimes the output resolution is the only specified resolution, giving the impression that you can also scan at this resolution, which is in fact not the case. As already mentioned, the raster function and the corresponding high output resolution for greyscale pictures are not required. It is also important to know that high scanning resolutions are only essentially required for drawings and not for greyscale pictures. This point will be elaborated further in later chapters.

Let us consider whether using a higher output resolution for drawings is sensible. What the scanner hardware does not recognise at 300 or 400 dpi may not be used – or may it? First, however, let us say that the high-resolution specification of many scanners is of no use, since it is only relevant to the output resolution of rasterised pictures, which in any case is not needed. In line drawings it is of no extra value if a scanner can select output resolutions of more than 300 or 400 dpi, if only the size of the data set is increased. The accuracy is not increased if the physical resolution of the scanner is only 300 or 400 dpi.

However, scanners (or scanner software) do exist which, by innovation, calculate a higher resolution and also attain a higher accuracy. When scanning drawings, if a scanning resolution of 800 dpi, for example, is selected these scanners automatically switch to greyscale scanning. As the scanner cannot offer resolutions higher than 300 or 400 dpi, the drawing is not scanned at a resolution of 800 dpi but, in the same way as a photograph, as a greyscale picture using a resolution of 300 or 400 dpi. Hence no longer are only black and white values processed, but instead grey levels with, for example, 256 levels of brightness are read. From this grey-level picture the scanner software then produces a black and white drawing with an output resolution of 800 dpi. The software can produce more accurate lines from information which is a factor of 256 times larger, i.e. the lines are not frayed and stepped, but smooth and even. However, it is important to know what physical resolution these calculations are based upon. There is a big difference in whether a scanner has a 300 dpi or 400 dpi resolution, as there is a difference of 33% here.

Lenses, mirrors and sources of light

In addition, the light source, mirrors and lenses comprising a scanner system affect its quality. Such a system should be mounted rigidly on to a slide and the individual parts must not move with respect to one another. If the mirrors are not connected rigidly to the lenses, but are forced to follow them during scanning, the focal length may possibly be altered. This can lead to distortion.

As colour photographs are of course also scanned using greyscale scanners and then reproduced as black and white pictures, the light source within the scanner also plays an important role. A scanner is only able to recognise all the colours of an original document if the light source used transmits the complete spectrum of visible light. As a result the ideal light source should transmit white light, which contains the complete spectrum of light from violet and blue through green and yellow to orange and red. White light is therefore a mixture of coloured light.

For example, if green LEDs are used as the light source, a particular spectrum of light is missing from this source. Hence not all the colours in an original document can be scanned equally. These incomplete colours are known as missing or drop-out colours. The drop-out colours should be mentioned in the technical data sheet for the scanner. Unfortunately, data sheets are often incomplete in this respect. Using the naked eye a light source may be examined only with difficulty, since even when the eye considers a light source to be 'white', a particular light spectrum may be missing or it may only be transmitted very weakly. For example, a white fluorescent light may have the drop-out colour of yellow.

A diverse range of light sources exist with various characteristics. A halogen lamp is relatively free of problems and has an optimum and smooth spectral light distribution as well as a high light yield. The disadvantage of a halogen lamp is the large amount of heat generated, so that when used in a scanner a fan needs to be included. Many types of fluorescent lights exist with various characteristics, i.e. the spectral light distribution and the light yield may vary significantly between different models. Therefore, if you are using a fluorescent light its actual light spectrum and light yield must be taken carefully into account.

A greyscale scanner must be able to recognise all the colours in an original document as effectively as it recognises grey levels. A colour scanner on the other hand must separate the individual colours (red, green and blue) and save them separately. Various processes exist for this. A colour scanner may, for example, contain only one light source and so have to scan an original document three times, with a different filter (red, green, blue) being connected between the light source and the document each time. On the other hand three light sources (red, green, blue) may be present, whereby scanning needs only one pass, during which the three lights continually but alternately illuminate momentarily. These three lights must also be able to deal with violet, yellow and orange.

Other scanners exist which have only one light but still scan an original document only once. These scanners operate using either a rotating colour filter or three CCD lines.

Care must be taken when using a colour scanner to ensure that it is capable of producing a wide spectrum of light, even when operating in black and white mode. If a colour scanner has three light sources (red, green, blue), but uses only the green light when scanning in black and white mode, then all the colours in the original document may not be converted by the same degree into grey levels. This also applies to colour scanners which scan using one white light source and a colour filter system. This is the case with scanners where the development engineers have 'forgotten' that besides the red, green and blue filter settings an additional neutral setting (no filter) is required for greyscale scanning. Therefore, a comprehensive scanner data sheet should mention which type of light is used as well as what spectrum of light can be attained.

When a so-called transparency adapter is attached to a scanner, slides and negatives may also be scanned. The resolution of the scanner must be suitably high for the smaller format. In an transparency adapter the slide is not illuminated with the reflected light being measured, but instead the light shines through the slide, as in a slide projector, on to the measuring unit of the scanner.

Lines and distortion in a picture

As the operation of electrical components changes with age, individual CCD sensors, for example, could report a different measurement for the same grey level. The light source may also age during its service life. For this reason each time a scanning process is started the scanning device should be calibrated for black and white values (e.g. by scanning reference areas). The values measured are used as control values for 'black' and 'white'. Therefore, errors due to a varying light source or to a sensor operating incorrectly are compensated for. In addition, the maximum dynamic response range of CCD sensors may then be utilised fully. If a scanner is not internally calibrated, lines may appear in greyscale pictures due to sensors not operating uniformly. Errors here have more effect on black than on white areas.

Whilst the sensor slide moves slowly during scanning, thousands of picture dots need to be processed. When a 300 dpi scanner is used each line produces 2550 picture dots. Here it is important that when the sensors move forward, the movement is uniform. As the transfer of data from the scanner to the computer is not usually fast enough, most scanners need to pause intermediately when scanning. A slow hard disk may also be the cause of data not being transferred quickly enough to the computer. This means that complex and time-consuming control electronics need to be used. The slide needs to be repeatedly accelerated, decelerated, reversed minutely and then continued. If a scanner does not backstep after a pause, the scanning velocity is too low on continuation. This may lead to distortion especially when reading line drawings and text and which, when using a text recognition program (OCR), for example, may lead to a very high failure rate.

However, an elegant way of bypassing the 'stop and go' problem when moving the sensors does exist. If a scanner has an integrated buffer memory of 1 Mbyte or higher, the data is then stored intermediately in the scanner and the sensors can move smoothly without stopping. If during scanning, data is already transferred to the computer, the buffer memory is not filled so quickly. However, even here the pausing of sensor movement must be controlled by complex and time-consuming means since pictures requiring more than 1 Mbyte of memory must be possible to scan. Furthermore, there are scanners available in which the scanning velocity may be preset. Here, if scanning at a high resolution, the velocity can be decreased to a level where the scanning process is not interrupted.

Scanner software

Much discussion takes place on scanner hardware. Of course the technical data is important, especially that not found in the manufacturer's data sheets, but the software is at least as important. A scanned picture is of little value if it cannot be printed. Hence a scanner needs to be able to take into account the characteristics of the output device. In addition, the characteristics of the original pictures need attention. Not only photographs, but also pictures from magazines and catalogues need to be scanned. As printed pictures are rasterised, a scanner must be capable of 'derasterising', or else undesirable moiré effects (interference patterns) are obtained. Also, the quality of line drawings may be considerably improved by means of software, in that the scanner internally scans line drawings using the greyscale mode and then calculates the corresponding black and white values.

Some types of scanner software, however, still leave much to be desired as far as basic functions are concerned. For example, it is frustrating when it is not possible to enlarge or reduce the scanned picture by various magnitudes. If the so-called 'pre-scan' function is also missing, then this is even more annoying. With a good pre-scan function a complete page may be scanned at low resolution, and a section subsequently selected, enlarged and then scanned at high resolution.

Optimum scanner software does not need resolution, brightness, contrast and gradation to be calculated and input, but instead only the desired result (e.g. type of picture, quality of picture, size of picture) need be specified. The scanner then calculates all the settings itself. With the aid of a few standard settings and without

manual image processing a good-quality picture can still be produced. These standard settings must be selectable by some simple means. In addition, an optimum software package should enable a greyscale picture to be displayed on the monitor with at least 64 grey levels. Only then may a picture be properly assessed. The quality of a scanner is also dependent to a large degree on the software.

Types of scanners

Before completing this chapter we will quickly review various types of scanners. The most common scanner is the flat-bed scanner. Here an original document is placed on to a glass plate and a slide with sensors passes over it during the reading process. Another type of scanner, not in great use nowadays, is the feeder scanner which draws pages one by one through itself. In these scanners, the paper moves whilst the sensors are fixed rigidly in place. As original documents for these scanners must always consist of individual sheets, implying that scanning books is not possible, they are rarely used. Furthermore, an automatic sheet feeder is available for flat-bed scanners.

If scanning needs to take place at a very high resolution (e.g. with slides, negatives, small line drawings), a so-called camera scanner may be used. As with an ordinary camera, a picture is taken with a camera scanner via an objective lens. However, CCD sensors are used instead of film. The camera is positioned vertically above the original document on a tripod. In professional camera scanners the distance from the camera to the surface to be scanned may be changed and hence the resolution with it. The resolution may be increased considerably for small original documents. For example, a resolution of 600 dpi may be attained for an A4 original and a resolution of 1000 dpi for an A5 original.

Another advantage of the camera scanner is that, by using an transparency adapter, slides, negatives and even objects may be scanned. The problem which occurs when scanning slides and negatives is that the picture needs to be considerably enlarged. Therefore, the physical resolution of the scanner needs to be high. Better results are obtained for the dark areas of pictures in slides and negatives. A prerequisite is that the physical resolution of the scanner be sufficiently high and a so-called transparency adapter be available. Here the light is not transmitted by the scanner and reflected by the slide, but is instead shone directly through the slide, as in a slide projector, to the scanning unit of the scanner. With such scanners it is even possible, for example, to scan X-rays and similar material. The advantages need to be weighed against several undesirable effects. Camera scanners are difficult to use, the original document is not enclosed and there is no screening on the scanning surface, so that reflections due to different light sources may be produced. Camera scanners also require more room for the tripod and lights. In addition, with some camera scanners, the associated software leaves a lot to be desired.

Summary

The subject of scanning covers a wide range of processes bordering on photosetting, printing and visual impact. As the complex relationships between these processes are insufficiently considered, even by manufacturers and sales people, a DTP user often stands alone. Even the scanner hardware and software is still generally found to be severely lacking. However, there are essentially relatively few matters which need to be taken into account and few aids are required to attain good-quality pictures.

In order to describe more precisely the source of individual problems and how they affect each other, the most important characteristics of scanners will be dealt with in more detail at other relevant places in this book. However, it is not important to understand everything down to the last minute detail. The emphasis is more on understanding the basic relationships, so that practical tips and advice may be used to an optimum degree.

Many scanner sales people, for example, quote the very high resolutions that their scanners possess. This usually has little to do with reality and probably serves to promote sales rather than having any useful meaning. Well, it is simpler to say that a scanner has a resolution of 1500 dpi rather than to be concerned about complex relationships. It is even possible to astound a scanner sales person completely if you demonstrate that a resolution of about 150 dpi is often sufficient for professional image processing.

However, what is undesirable, to say the least, is that customers are often not informed of the physical (optical) resolution of the scanner. As already explained, standard flat-bed scanners presently have a maximum physical resolution on the *x* axis of 400 dpi. Flat-bed scanners with a resolution of more than 400 dpi are still very expensive. However, some scanners may achieve higher resolutions when scanning drawings via extrapolation.

Even more undesirable is the fact that scanner software exists which allows a high resolution to be entered for grey-level picture scanning leading to the false assumption that the scanner will be scanning at this resolution. In reality, however, the output resolution is entered and simply a pixel structure using this resolution is created. The scanner is therefore scanning at a much lower resolution then was entered through the software. For example, when producing a photograph from a

photosetter at a resolution of 2540 dpi a scanner resolution of about 200 dpi is all that is required. As original pictures are usually larger than the required picture, often a much lower scanning resolution may be used.

High scanner resolutions are only required for high-quality line drawings or for small originals. In any case, with drawings it is more sensible to use larger originals and thus achieve an increase in the resolution indirectly.

More important than the resolution is the quality of the scanner, the capability of the software and correct operation. Here the primary issues are those which have so far not been mentioned sufficiently in data sheets. For example, the quality of the software determines whether it is sensible to scan a drawing using a resolution of more than 400 dpi, or whether it is possible to scan rasterised pictures (e.g. postcards or pictures from magazines) by some simple means.

What can be achieved using mature scanning and image processing software will be described in more detail later. However, it will also be made clear that certain aspects are still in need of improvement.

Chapter 4

Obtaining text and archiving

ABC

Text recognition (OCR)

Is it really necessary to retype text into the computer which has already been typed or printed before? So-called OCR programs promise to assist here. 'OCR' stands for 'Optical Character Recognition'. With the aid of a program and a scanner original text does not need to be retyped. Nearly all scanners today may be used in conjunction with an OCR program to acquire text.

Perhaps you can still remember that learning to read was not so easy for all of us. This is also true of OCR programs, which must first learn to recognise the shape of characters. Nowadays OCR programs can read almost any typeface, not only the so-called OCR fonts. OCR fonts are, for example, found at the foot of cheques or on transfer forms.

How does an OCR program work? As you already know, the scanner scans the original image point by point and transfers coded information to the computer. Essentially all that exists is a large number of black and white dots. If letters of the alphabet are scanned, they are not recognised as such by the computer but instead are regarded as pictures of letters, comprising a large number of individual pixels. Such pictures of letters cannot be processed further using a word processing program but rather only by a drawing program and this is of little use.

A word processing program does not take note of the individual pixels as a drawing program does, but instead only considers the code representing each individual letter. Hence a word processing program does not send pixel information to the printer, but instead merely transmits the codes for the individual letters. For example, the letter 'A' is coded by the number 65, known more fully as ASCII character 65. If the printer receives ASCII character 65, it determines from an internal table (ASCII table) that an 'A' needs to be printed.

An OCR program therefore has the task of systematically comparing the dot pattern read in by the scanner with a list of stored pictures of letters held permanently in the program and then saving the corresponding code for the matching letter. Once a letter is found, the search for the next letter picture is started again from scratch.

Various processes exist to evaluate the picture pixel structure in order to recognise letters. The simplest process is to enlarge or reduce the unrecognised pixel picture to a fixed size and then to compare it pixel by pixel with stored originals. The number of pixels which exactly match those in an original is a measure of how accurately the letter is recognised. However, the problem is that hardly any letter will coincide exactly in every respect with a stored original. Another OCR method is vectorisation where characters are recognised by their characteristic features. Good programs combine several methods in order to ensure that characters are recognised as easily and accurately as possible.

The main difficulty in evaluating pixel patterns is due to the wide range of completely different fonts. Even proportionally spaced fonts in justified text causes problems, as in this case each character assumes a different width. In addition, there are text attributes such as bold, italic and underlines to contend with. How can a program differentiate between the two letters 'rn' and the single 'm'? It is also naturally a problem when the quality of the original document being scanned is poor.

Programs exist which are capable of learning whilst other programs know 'everything' to start with and do not want to learn any more. If a program which recognises 15 fonts is faced with a 16th unknown font, the final result may not be especially good. Systems which are capable of learning have certain advantages, since special characters, or characters which are only poorly defined, may then be recognised and if necessary corrected.

A program should also be fast. Peak speeds of 800 characters per second achievable by professional systems are definitely not reached with a standard OCR program. However, you already have a fairly fast program operating at about 50 characters per second. It is also important to know how many character patterns the program can use in order to learn further fonts and special characters.

It should now be clear why OCR programs are only suitable for certain fonts and why they should be 'capable of learning'. If an OCR program does not recognise a certain picture of a character, it is possible to inform the program which character the picture represents and the program will then remember it for the future. However, unassisted learning capability is a long way from determining how accurately an original text can be read. It is possible for a program which is not capable of learning to outperform completely a program which is.

ABC

As no OCR program currently operates completely error free, rechecking the text with a spell check program and manual checking cannot be avoided. The error rate depends on the quality of the OCR program, the quality of the original font and also the resolution of the scanner. Scanners with 400 dpi resolution produce better results than scanners with 300 dpi.

In spite of several sources of errors good OCR programs are a considerable help and lead to a considerable saving in time. However, good typists are preferable to simple programs.

Archiving of documents

Each day the mountain of paper grows, demanding to be administered. The documents needing attention require ever increasing resources of people, material and space. Would it not be more sensible to record all the documents to be archived by means of a scanner and then to administer them electronically?

With suitable archiving programs the electronic administration of documents is no longer a problem today. The saving made is readily apparent if you consider that the complete contents of a filing cabinet containing about 30 files can be saved on a single optical storage disk. Once the data has been saved, a document may quickly be found again by means of powerful search mechanisms.

Let us first consider the recording of documents. Using a scanner with an automatic sheet feeder, individual pages can be scanned as pixel graphics. Company logos, notes in margins, date-received stamps, signatures and other information are retained. Since an A4 sheet comprises about 2 Mbytes of memory, the data is compressed before it is saved, so that a file may be reduced to about 500 kbytes. With suitable hardware the complete process of scanning, compression and storage can take place very quickly.

During the recording of information search keys need to be fixed, through which the document may be found again. Existing filing systems may be maintained for this. Complete sentences, certain individual words, numbers or dates may be used as search keys. Because the complete page is saved simply as a pixel graphics drawing, the computer can only search for specific contents when suitable search keys have been associated with a particular document beforehand. To make matters easier, an integrated text recognition system is used to convert pixel pictures to text. Addresses, salient words or complete passages of text may be stored automatically as search keys by using such a system.

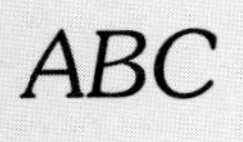

Any on-line computer may search and view the documents stored in these electronic databases. The complete page is displayed true to the original on the screen. Hence, several office departments may quickly view documents at the same time. If someone requires the document on paper, they can request to have it printed. An interesting possibility is so-called clipping, whereby a complete new page can be created from several documents. Interesting information is copied and added to a new document which is then given new search keys. It is therefore even easier to find your way around the huge mountain of information.

The high-volume write-once optical storage devices are usually used for storage, i.e. the so-called WORM cartridges (WORM = Write Once, Read Multiple). As data is 'burnt-in' via a laser beam, these storage devices can only be written to once so information cannot be deleted by mistake or changed. This type of storage conforms to industrial and governmental regulations and may also be used to store sensitive company documents. The information burnt-in via laser beam is regarded as valid protection against fraud. By using access rights and user passwords access by unauthorised persons may be denied.

For example, 800 Mbytes of data may be stored on a WORM cartridge. This is sufficient for about 16 000 A4 pages with a resolution of 400 dpi. This capacity is based upon an average compression ratio of 1:40. The compression factor is dependent on the proportion of black regions, as opposed to white, in a document. In order to expand the capacity even more, cartridge units are available which provide uniform access to multiple cartridges.

The implementation of document archiving facilities may expand as the requirements grow. Hence, you can start with a 'single seat' system and then expand to department and company models. By means of a network all personnel with authorised access can be connected to a central database.

Chapter 5
Scanning drawings

General

Pictures are not all the same. A strict differentiation must be made between line drawings, photographs and rasterised images. In addition, we must deal with the often heard presumption that scanner resolutions are just about adequate for line drawings, but far too low for greyscale pictures. The exact opposite is true! A high-quality photograph may be output at a scanner resolution of only 150 dpi. However, we will consider this again in the next chapter.

With pictures you must basically consider the memory requirements and the computer time required by the output device. This means that you should only scan at the resolution that is necessary. If an original document is already of poor quality, scanning at a high resolution does not help.

In line drawings, the scanner need only recognise two states: black and white. Accordingly, the scanner needs to be set to the 'black and white' mode (line type). The only parameters which need to be entered are the resolution and the limit (brightness). This limit determines what minimum grey level the scanner should use to recognise a dot as black. Depending on the original document (fine or thick lines, deep black or grey lines) a different value may be entered for this limit.

Scanning resolution for drawings

The resolution is a measure of the accuracy with which the lines are reproduced. When scanning drawings the scanning resolution may be selected according to the capabilities of the scanner and with respect to the resolution of the output device. A resolution which is too high produces unnecessary costs in the photosetting process, as in general photosetting companies calculate cost according to the amount of memory used or amount of time taken for photosetting. On one hand the scanning resolution should not be much higher than the resolution of the output device.

Hence it is not very sensible to scan at a resolution of 600 dpi if printing is done on a 300 dpi laser printer. On the other hand, due to the construction of the scanner (physical resolution), a resolution which is too high does not necessarily produce a corresponding improvement in quality in every case.

➪ *By selecting the correct scanning resolution, cost and memory requirements may be reduced with no loss in quality.*

You should be aware that a 1500 dpi scanner usually possesses a 'real' physical resolution of only 300 or 400 dpi. Some scanners have interpolation capabilities which can produce a far better result with an output resolution of more than 800 dpi by interpolation of a 400 dpi scanned image. The lines are smoother and do not appear jagged. This interpolation process does, however, have limits where extremely fine lines, or lines lying close together, need to be recognised. When printed these lines appear marginally thicker and may therefore merge.

Scanners also exist where a scanning resolution higher than the physical resolution may be set. However, no improvement in quality is achieved. Hence scanning at a resolution higher than the physical resolution is basically not sensible. If your scanner, for example, allows you to set a resolution of more than 300 or 400 dpi, this does not automatically mean that you will achieve better results.

Which scanner resolutions are sensible depends not only on the physical resolution, but also on the quality of the scanner software. In order to evaluate your scanner software, you should scan various original documents with different settings and then have these photoset. During these tests, you should vary not only the resolution but also the limit (brightness). An enlarged representation on the screen usually results in lines being scanned too thinly. Only when the photoset printout has been examined can the scanning be properly evaluated. Therefore, gain experience before you scan hundreds of pictures for your next catalogue. First you should enquire as to what physical resolution the scanner hardware possesses on the *x* and *y* axes (provided your scanner sales person knows the correct answer!). If you are informed that you have a resolution of more than 300 or 400 dpi, you should sit up and take notice!

For critical drawings a larger original should be used, which can then be reduced through the scanner software. The resolution of the picture increases by the same factor by which the drawing is reduced. If, for example, a 400 dpi scanned drawing is then reduced by a factor of five via the DTP program, an image resolution of 2000 dpi is attained. Poor-quality line drawings (light lines) may often be improved using a good photocopier.

In general, with image resolutions of about 600 or 800 dpi excellent quality may be achieved. An unnecessarily high scanning resolution does not bring about an improvement in quality and requires considerable memory. By doubling the resolution, the memory requirement quadruples. The quality of scanned line drawings may to a certain degree be evaluated on the screen. To achieve this the picture needs to be enlarged (some scanner software can only show the picture enlarged) so that all lines can be checked. On one hand lines must not be too thick, otherwise they will merge if they are close together, but on the other hand thin lines must be completely visible and not have any breaks. The reduced representation on the screen cannot be used to evaluate quality, as the resolution of the screen is not sufficient for an exact representation. A realistic check of image quality is therefore only possible once photosetting has taken place.

Image size and scanning resolution

As already mentioned, the final image size must be taken into account when scanning. If, for example, a picture is scanned at 300 dpi and then reduced by half in the DTP software, the resolution of the picture increases accordingly to 600 dpi. If a picture is to be reduced, then the scanning resolution can be reduced. If it is to be enlarged, the scanning resolution can be increased. The scanning resolution is usually effected by one of two alternatives: either it may be suitably changed or the image size may be entered in the scanner software.

❒ Changing the scanning resolution

If the image size cannot be entered in the scanner software or if the image size is set at 100%, the required scanning resolution is calculated as follows:

$$\textit{Scanning resolution} = \textit{required image resolution} \times \frac{\textit{desired size}}{\textit{size of original}}$$

If an image resolution of 600 dpi is required and if the length of the original picture is to be reduced from 16 to 10 cm, the scanning resolution is calculated as follows: 600 dpi × 10 cm/16 cm = 375 dpi.

❒ Entering the image size

If both the desired image resolution and the image size are entered to define the scanning resolution, then the internal scanning resolution is correspondingly changed. If an image size of more than 100% is entered, then the resolution is increased; if less than 100% the resolution is decreased. Once the image size is

entered, the scanning resolution does not have to be calculated separately. The following is therefore true: scanning resolution = desired image resolution. The image size to be set can be calculated as follows:

$$\textit{Image size in \%} = \frac{\textit{desired size}}{\textit{size of original}} \times 100$$

If, for example, the length of the original is 16 cm and it is to be reduced to 10 cm, the image size can be calculated as follows: 10 cm/16 cm × 100 = 62.5%.

By entering the size of the image, it is possible to save the time spent on calculation. If, for example, the picture is to be printed on a laser printer, the scanning resolution may be set to the printer resolution of 300 dpi, for example. If the desired image size is now entered, the internal scanner resolution is changed in such a way that the reduced or enlarged picture again has a resolution of 300 dpi.

The actual scanning resolution resulting from the image size being taken into account is known as the effective scanning resolution. The effective resolution is thus not the resolution which is set, but the resolution at which scanning actually takes place. When an image size of 100% is set, the set resolution is of course equal to the effective resolution. Suppose a scanning resolution of 300 dpi and a desired image size of 200% are entered; then in reality (effective scanning resolution) scanning takes place at 600 dpi (200% resolution of 300 dpi = 600 dpi). As a result the enlarged picture once again attains the desired resolution (of the picture) of 300 dpi. The effective scanning resolution may be calculated as follows:

$$\textit{Effective scanning resolution} = \textit{set scanning resolution} \times \frac{\textit{set image size (in\%)}}{100}$$

If an image size other than 100% is entered in the scanner software, the above formula can be used to determine at which resolution the scanner is actually operating (effective scanning resolution). If, for example, a resolution of 800 dpi and an image size of 50% is entered in the scanner software, the scanner scans at the following effective resolution: 800 × 50/100 = 400 dpi.

Raster/vector conversion (auto-tracing)

Next we will consider the fantastic possibility of increasing the picture quality to any level and yet saving storage space at the same time, i.e. raster to vector conversion. In pixel pictures each pixel is individually stored. In vector pictures, on the other hand, only formulae for the picture are stored. For example, only the starting and end points of a straight line need to be stored.

Conversion programs convert individual pixels into vectors. Conversion may take place either semi-automatically and interactively on the screen or fully automatically. Semi-automatic conversion takes place when you click the mouse on a curve. Hence you click and convert each line one at a time. The desired accuracy may be fixed separately for each element. If the conversion is fully automatic, the complete scanned drawing is converted into vectors to a predetermined accuracy.

When choosing a raster to vector conversion program you should ensure that it is simple to operate and enables vectorised drawings to be processed further. What use is a conversion program if further processing requires, in addition, an expensive CAD program which is difficult to use? Care should also be taken that the conversion program fixes as few anchor points as possible, so that further processing is simple. Also curves and circles should not be formed from individual straight lines.

Furthermore, conversion is usually only sensible if drawings are to be photoset. Care must therefore be taken that the conversion program produces data which may be photoset. If a converted curve is too complex, the drawing may be printed using a laser printer, but the photosetter will refuse this work. For this reason a conversion program should undergo the so-called 'limit check'. This check examines how many anchor points a line comprises. If the number of anchor points is too high, the accuracy for this line needs to be reduced a little.

However, before conversion can take place the drawing must first have been obtained using a scanner. Here the highest possible scanning resolution should always be used for scanning.

Pixels and vectors

In order to produce good DTP results, a basic understanding of the general relationships is important. This includes understanding the pixel or vector orientation of picture data. For this reason we would like to add a few lines on this subject:

If you scan a picture (photograph or line drawing), the original is scanned by the scanner dot by dot. For each dot, information is stored specifying its brightness. With a line drawing it is merely decided whether each dot is black or white. It does not make any difference to the scanner whether the original is a piece of text or a picture, i.e. even text is handled as a picture. Hence scanned text cannot be further processed immediately using a word processing program. Instead, it may only be dealt with as a picture using a drawing program.

In order to process text with a word processing program, so-called OCR programs (text recognition programs) need to be used. The text recognition program searches for pictures which appear to be letters and identifies them. The information is no longer held in black and white pixels, but instead as characters such as the letter A or B (ASCII characters).

However, let us return to the pixel picture. You have a scanned picture which consists of black and white pixels (dots). First enlarge this picture. Now rearrange the individual pixels with a larger gap between adjacent ones. The number of pixels, however, remains constant and hence the quality of the picture naturally deteriorates. This type of picture is known as pixel orientated.

There are, for example, drawing programs which operate in either pixel-orientated or vector-orientated mode. What is the difference? As already explained, in a pixel-orientated picture a drawing is made up of individual dots. In a vector-orientated picture this is not the case. Here the individual vectors are stored, i.e. for each line the information available defines where this line starts and where it finishes. In order to change the size of the picture, only the dimensions need to be changed and the quality of the picture does not deteriorate. The resolution is limited simply by the output device.

The advantage of vector-orientated drawings is that less memory is required and pictures of very high quality can be photoset. In addition, pictures may be enlarged without loss of quality. An example of a vector-orientated drawing program is GEM Artline.

The conversion of pixel pictures (raster pictures) from the scanner into vector-orientated pictures is not without its problems. For example, when a circle is converted it should not be formed of short straight lines. Care must also be taken that the converted data can still be photoset. If a line consists of too many anchor points, the photosetter cannot process it further.

Rudimentary conversion programs do not automatically duplicate varying line thicknesses and filled areas, but simply reproduce the outlines of the drawing.

Chapter 6
Scanning photographs

Now we will consider the scanning of photographs. However, before we can start properly, we need to explain a few more essentials. Hence in this chapter the main problems of scanning and the reproduction of photographs will be described in detail.

To help you orientate yourself, the list below shows how the subject area of 'Scanning and printing of photographs' is distributed over the following chapters:

- Chapter 6 'Scanning photographs'

  Here the basic problems are described. The most important subject areas covered are raster, grey levels and scanning resolution.

- Chapter 7 'Scanning rasterised pictures'

  Described here is what must be taken into account when a printed picture (e.g. from a brochure) and not a photograph is to be scanned. These notes should help you avoid the so-called 'moiré effect' (interference patterns in the picture).

- Chapter 8 'Camera-ready copy and photosetting'

  This chapter describes the way in which professional camera-ready copy is created. What DTP users need to consider when they provide data for a photosetter will be explained. However, there is also advice for photosetting companies.

- Chapter 9 'Printing'

  This is a very important chapter. Here the problems which occur during reproduction because of offset printing are described in detail. If you are of the opinion that these problems are only of interest to printers you will appreciate that issues are not quite that straightforward. The printing process needs to be taken into account as soon as image processing starts.

- Chapter 10 'Processing greyscale pictures'

  Here it will be explained how greyscale pictures should be processed and prepared for printing. An additional important aspect is the calibration of the monitor. Tips on image processing will also be given.

- ❐ Chapter 11 'Pictures from video cameras'

  It will be shown here how a video camera, instead of a scanner, may be used to deliver pictures.

- ❐ Chapter 12 'Scanning and printing of pictures: summary and tables'

  This chapter completes the subject area of 'Scanning and printing of pictures'. In short, everything which needs to be known is clearly integrated. Furthermore, all the tables which are important for image processing are included here. If you do not consider the problems and relationships described, you will not obtain optimum results. As you will see, the task is not complete merely by scanning the photographs. The pictures also need to be printed and the real problems begin here.

Before you start scanning and processing many pictures, a few more basic principles must be described. In this chapter you will discover that your scanner has a higher resolution than necessary for scanning photographs. The most important factor which determines the quality of your pictures is the number of grey levels which your scanner is able to recognise.

Grey levels cannot be printed!

Please excuse this minor hair-splitting remark, but it is true: grey levels cannot be printed. You will not find a single grey dot in any printed book, just black or coloured dots.

How is a grey level simulated?

As no real grey levels can be printed, a grey tone needs to be simulated. If a sheet of paper is filled with dots and then viewed at a distance, the dots blur and become a grey area, i.e. a grey level is simulated. In reality only black dots are present. This is known here as a rasterised picture.

If this grey area is to be made darker, the individual dots need to be a little larger, so that the white area remaining becomes smaller. From a distance, the individual larger dots are no longer seen, but instead all that is perceived is an area which is a little darker.

Hence it is possible to simulate grey levels with various sizes of black dots. The effect is due to white areas being reduced or increased, so that the total area appears darker or lighter.

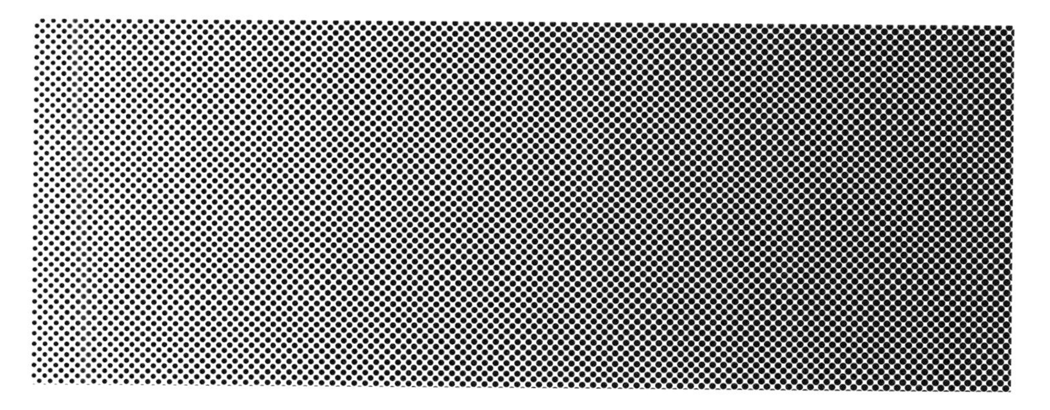

Owing to the various sizes of the raster dots different grey levels are simulated.

What is the structure of a printed picture?

A picture comprises dots of various sizes whereby, due to an increase or reduction in the white area between individual dots, light and dark patches are created. The individual dots are known as raster dots. The more dots which are used, the smaller they are and the better the quality of the picture.

Fotograph: Jutta Hofmann

A picture is built up of varying sizes of raster dots. Owing to the remaining area of white, different grey levels are simulated.

What is a raster?

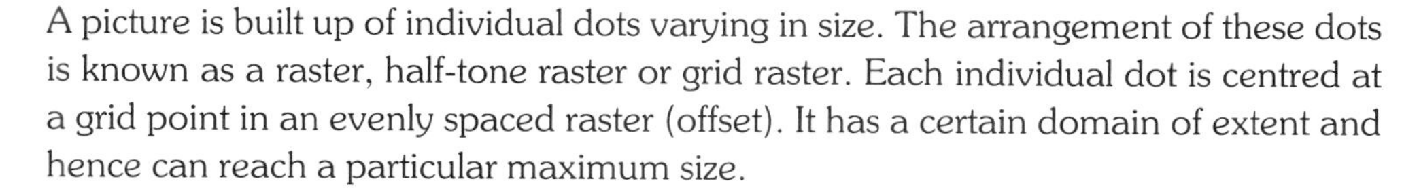

A picture is built up of individual dots varying in size. The arrangement of these dots is known as a raster, half-tone raster or grid raster. Each individual dot is centred at a grid point in an evenly spaced raster (offset). It has a certain domain of extent and hence can reach a particular maximum size.

The number of dots per centimetre or per inch is known as the raster width, frequency of the half-tone raster or frequency of the grid raster. The more dots used, the more closely meshed is the grid and hence the dots are smaller, improving the quality of the picture.

Dimension for raster width

The raster is specified with respect to lines and not areas, i.e. it is specified by how many lines are used per centimetre or per inch, or by how many raster dots are used on a line per centimetre or per inch. Hence the term used is 'lines per centimetre (L/cm)' or 'lines per inch (lpi)'. A raster with 60 raster dots per centimetre is equivalent to 60 L/cm or 150 lpi. To convert from L/cm into lpi see the next section.

Conversion of centimetres into inches

As values in DTP are usually given in inches and not centimetres, they may need to be converted. One inch equals 2.54 cm:

- Length in inches = length in cm/2.54
- Length in cm = length in inches × 2.54

As the resolution is given by the number of pixels per centimetre or per inch, the conversion for resolution is the opposite of that in the above calculations for specifications of length:

- Resolution in inches = resolution in cm × 2.54
- Resolution in cm = resolution in inches/2.54

A raster with 60 dots per centimetre therefore produces 60 L/cm × 2.54 = 152 lpi. So, in the DTP program you do not enter the value 60, but 150 instead.

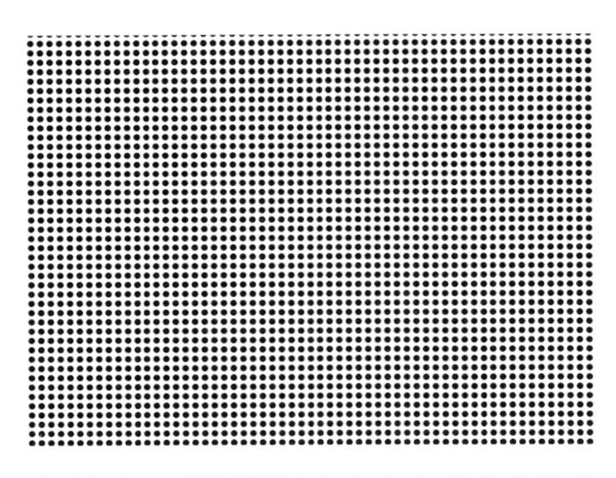

This area has a raster of 20 lpi (20 lines per inch) or 8 L/cm (8 lines per centimetre). The number of dots per centimetre or per inch determines the width of the raster. The larger the distance between the individual raster dots, the larger the dots and hence the coarser the picture.

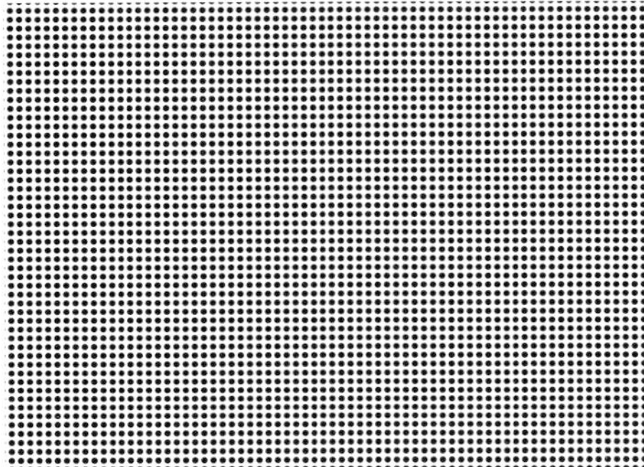

This area has a raster of 30 lpi (30 lines per inch) or 12 L/cm (12 lines per centimetre). The individual raster dots are separated by the same distance, i.e. they are arranged in a particular raster.

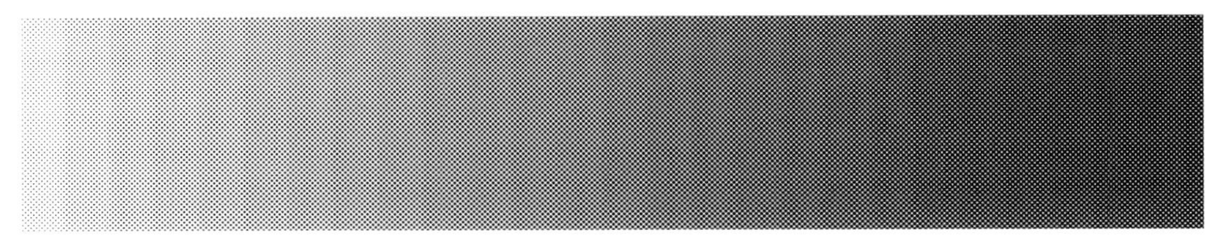

Each point has a particular area available for it. The larger the dot, the more white area covered and hence the darker the picture. This area has a raster of 60 lpi (60 lines per inch) or 24 L/cm (24 lines per centimetre).

This is the same area as above, but with a raster of 150 lpi or 60 L/cm.

Varying rasters –
pictures varying in quality.

Raster width:
75 lpi (30 L/cm)

Raster width:
120 lpi (47 L/cm)

Raster width:
150 lpi (60 L/cm)

All three pictures were photoset at 2540 dpi. The angle of the raster in each case was 45°.

It depends on the light: the printer cannot do it!

Why is it not possible to attain good-quality pictures using a standard 300 dpi laser printer and why must a professional scanner not have a resolution higher than 300 dpi? A picture with a raster of 60 L/cm (150 lpi) is already of very good quality. Newspapers, for example, only have a raster of 32 L/cm (81 lpi). Why is it not possible to achieve at least newspaper quality using a laser printer of 300 dpi resolution?

For the sake of simplicity we will mainly use DTP terms below. For example, for the raster of 60 L/cm the preferred term of 150 lpi will be used.

The printer cannot vary the size of pixels

As already described, grey levels are simulated by dots of varying sizes. But there is a snag: laser printers and photosetters can only generate pixels of identical size. Since the size of the pixels cannot be changed, what can be done?

The answer is simple. If two pixels are placed adjacent to each other, a larger dot is simulated. Add a few more pixels and the dot becomes even larger. Dots varying in size are therefore achieved by arranging several smaller pixels adjacent to each other.

For example, in an area of 5 × 5 pixels (5 area) 25 different sizes of dot may be attained and hence 25 grey levels. If 256 grey levels are to be created, each raster dot must be able to attain any of 256 different sizes. In this case a raster dot comprises an area of 16 × 16 pixels = 256 pixels of the same size, which arranged together produce a dot.

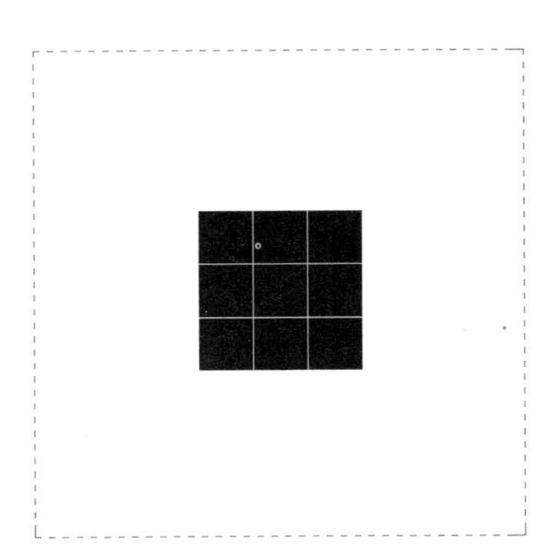

Laser printers or photosetters can only produce pixels of the same size. Dots of varying sizes are produced by grouping several pixels together. Individual pixels grouped together thus produce one raster dot. In the example here there are 9 pixels arranged in the raster pixel matrix as one raster dot.

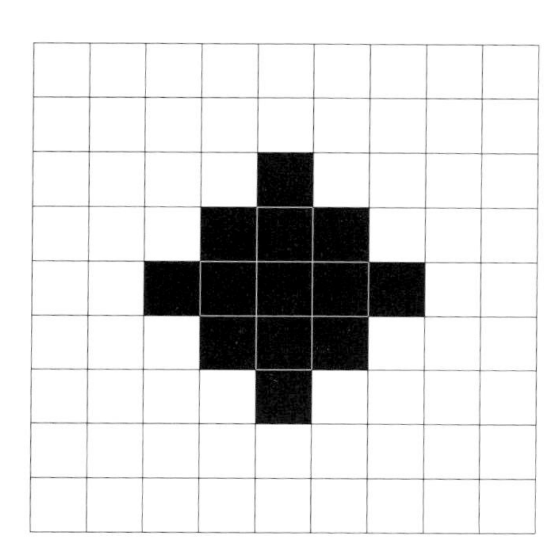

A raster dot is formed from a raster area, in which a specific number of pixels may be placed. The possible number of pixels in a raster matrix determines how many different sizes of dots can be produced and hence how many grey levels can be obtained.

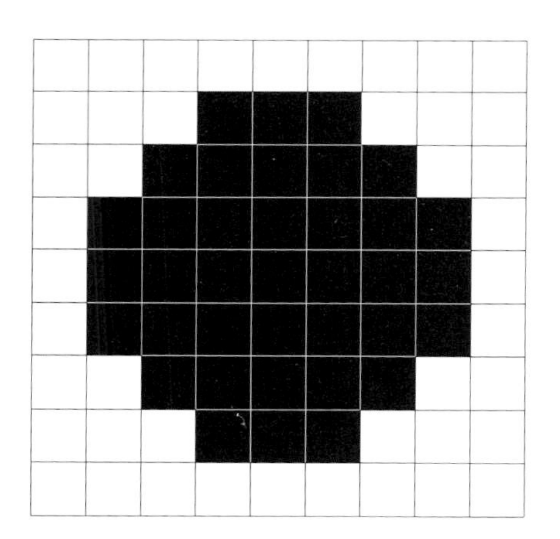

If more pixels are grouped together in the raster matrix, a larger raster dot is produced. Owing to the larger dot, less white raster area remains and hence that area in the picture becomes correspondingly darker.

This means that, for a picture with 25 grey levels, a raster area of 5 × 5 pixels must be reserved or, with respect to the line, a length of 5 pixels. If one dot is placed in the reserved raster area then a very light tone is attained, due to the remaining large area of white. If several pixels are added together, a smaller area of white is left due to the resultant dot being larger. Hence a darker tone of grey is achieved. The shape of the dots is in practice not the ideal one shown in the diagrams, as the laser beam from a printer or photosetter is not square, but circular.

How many raster dots can the laser printer print?

Let us now examine the laser printer a little closer. The laser printer has a resolution of 300 dpi. This means that 300 pixels may be placed per inch. In order to attain 25 grey levels, however, 5 pixels (with respect to the line) must be reserved for each raster dot as already mentioned.

If the resolution of the laser printer is divided by the number of pixels required for each raster dot, then the number of raster dots which the printer can output is produced. In this case 300 dpi/5 pixels = 60 lpi. The number of raster dots determines the raster width. The laser printer thus effectively only prints using 60 raster dots per inch (lines per inch) with 25 grey levels per raster dot. Converted into centimetres, this is: 60 lpi/2.54 = 24 L/cm (lines per centimetre); 24 L/cm is, however, lower than newspaper quality. We must explicitly point out that the above comments only illustrate the principal function where a raster angle of 0° was used. As will be explained in the following section, however, a raster angle of 45° is used in black and white pictures, from which a raster width of 53 lpi (21 L/cm) is then calculated. In Chapter 12 tables and formulae are given for possible raster widths.

Raster angle

As already explained, the picture quality is determined by the raster width and the number of grey levels thus produced. Accordingly the desired raster width needs to be entered into the DTP software for each picture. However, not only the raster width but also the raster angle need to be entered. What is the raster angle?

The raster angle is determined by the particular angle at which the individual raster dots are arranged. Its importance is due to the fact that the human eye can see horizontal and vertical lines particularly well. For this reason, a pattern of dots arranged at an angle of 0° is most easily recognised and hence tends to be distracting. The shape of the dots may also be changed, but we will come back to this later. Since it is not the raster itself, but rather the grey tones simulated by it which are to be recognised in a picture, a suitable raster angle is applied. The least recognisable raster pattern is that with a raster angle of 45°. In four-colour printing, a different raster width and angle is used for each colour to avoid moiré patterns.

 *In all single-colour printing the raster angle used is 45°.*

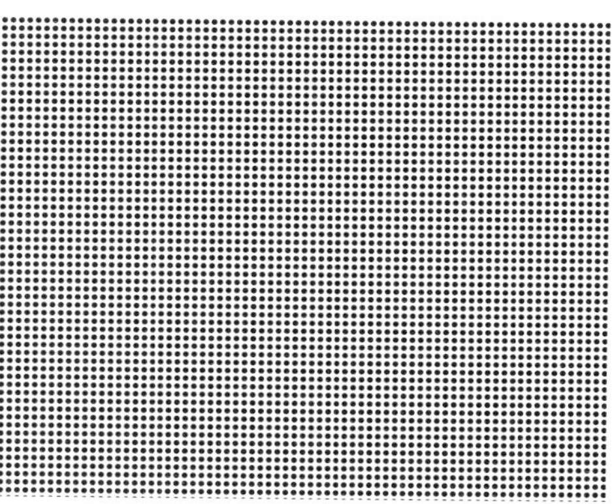

The left area has a raster angle of 45°, the right area has one of 0°. If you look at these areas from a distance, you will see that the pattern in the area on the right is more easily recognisable.

The left picture has a raster angle of 45°, the right one of 0°. The raster width is 54 lpi (21 L/cm) in both pictures. The difference can clearly be seen: the 45° angle is less distracting.

Here are the same pictures as on the previous page: left with a raster angle of 45° and right with one of 0°. Instead of using a raster width of 54 lpi these pictures were printed at 75 lpi (30 L/cm). This corresponds to a raster width found in daily newspapers.

Let us summarise briefly:

- One grey dot in a photograph is shown as one raster dot on printing.
- Raster dots of varying sizes, which are arranged at a fixed distance from each other (raster width) simulate the various grey levels.
- A raster dot comprises several pixels. With these pixels, dots of varying sizes are created. This process is used since printers and photosetters cannot normally vary the dot size.
- The terms raster width, half-tone raster, grid raster or frequency of a raster all have the same meaning. They determine the number of raster dots (not pixels) per centimetre or per inch. The dimension for raster width is L/cm (lines per centimetre) or lpi (lines per inch). The specifications for the raster width are with respect to the line and not the area.
- In DTP software the raster width and the raster angle need to be entered. In single-colour reproductions a raster angle of 45° is always used.

What are laser standard printers and scanners capable of?

The following examples should serve to show the principal capabilities of a printer and scanner. How the required scanning resolution is calculated will be described in detail at the end of this chapter.

The printer

You can now calculate what resolution the printer needs to output a picture with a raster of 155 lpi. This figure means that 155 raster dots per inch must be present. In order to obtain a good-quality picture, at least 64 grey levels must be produced. As already explained raster dots covering an area of 8 × 8 pixels are required in order to attain 64 grey levels. Since all specifications (resolution, raster) are given with respect to the line and not with respect to the area, 8 pixels per raster dot are required.

Thus 137 raster dots per inch (lpi) multiplied by 8 pixels per raster dot produces a total number of 1096 pixels per inch. The printer must therefore have a resolution of 1096 dpi in order to output a picture of 155 lpi with 64 grey levels. Unfortunately the laser printer is only capable of 300 dpi. This shows clearly that photographs cannot be output, even at newspaper quality, by means of a 300 dpi laser printer. Please note that these comments illustrate the principal function only and that a raster angle of 0° is used.

The Scanner

Further examples can be found in the following section. This time the required resolution for the scanner is to be calculated.

Suppose you would like to place an advertisement in a daily newspaper. In general photographs are printed in daily newspapers with a raster of 80 lpi. Hence 80 raster dots per inch are printed. Therefore only a little more than 80 grey levels per inch need to be obtained by the scanner, which means that only a resolution of about 80 dpi is necessary for scanning. Unbelievable, but true. If the picture is to be reduced, the scanning resolution may be even lower.

In order to acquire a 'feeling' for these problems, another example will be worked through. Suppose that you do not want your data to be photoset, but find the quality of the output from the laser printer acceptable. The required scanning resolution will now be calculated.

Since a raster width of only 53 lpi (21 L/cm) may be achieved using a 300 dpi laser printer, and for each raster dot only a little more than one grey level needs to be

scanned, scanning is sufficient at just over 53 dpi. In order to optimise the picture and for flexibility if the picture is to be slightly enlarged, the scanning resolution needs to be a little higher than the raster width.

The required scanning resolution is calculated as follows:

$$\textit{Scanning resolution} = \textit{raster width} \times 1.4$$
$$\textit{Scanning resolution} = 53\ \textit{lpi} \times 1.4 = 74\ \textit{dpi}$$

This means that a scanning resolution of more than 74 dpi for final output on a laser printer is not sensible. This conclusion is of course only valid when the printed picture is exactly the same size as the original.

Therefore, it does not make any sense if photographs destined to be printed on a laser printer are scanned at a high resolution, otherwise both a large amount of memory and a long print time would be necessary.

Why then are scanners with a resolution of more than 300 dpi necessary at all? High resolution is primarily required for line drawings as no grey levels need to be set here. In line drawings resolutions of 400 dpi and more are required.

The best-quality greyscale pictures are obtained if the scanned pictures are photoset. For this purpose scanner resolutions of about 200 dpi are required. However, if you wish to enlarge a picture, because the original is available only as a slide, for example, then high scanner resolutions are also required for grey-level pictures.

Instead of the factor 1.4 specified in the formula above, a factor of 1.2 may be used to reduce the amount of memory required. How this factor is derived and at which resolution scanning really should take place in practice will be reviewed in more detail later.

How many grey levels can be printed?

Top: 256 grey levels; bottom: 64 grey levels.

Of course, only the number of grey levels that the scanner is able to recognise may be printed. With 64 grey levels relatively good pictures may be obtained, but for perfect picture quality the scanner should be able to scan 256 real grey levels.

As already mentioned in detail, grey levels are produced by dots of varying sizes. In order to create these dots several pixels of the same size are grouped together. Hence if a picture file is photoset using a resolution of 2540 dpi (2540 pixels per inch) and a raster of 150 lpi (150 raster dots per inch), then for each raster dot the following number of pixels is present across the width of a raster dot: 2540 dpi/150 lpi = 16.

Length multiplied by length (16×16) gives the number of pixels comprising the raster dot area, i.e. 256 pixels for a raster dot and hence 256 grey levels. So the higher the printer resolution and the lower the raster width, the more grey levels which may be printed.

As the examples show, a very reasonable-quality picture may be obtained using

64 grey levels. The difference between this and using a 256 greyscale scanner is noticeable when dark areas on a picture are made lighter or when fine grey paths (especially in dark areas on a picture) are printed. Hence the subject and quality of the original also determine how clearly the difference between 64 and 256 grey levels is visible. In pictures with less than 64 grey levels, however, the loss in quality becomes very obvious.

The following formula may be used to calculate how many grey levels the laser printer or the photosetter can output for a certain raster width:

$$\textit{Number of grey levels} = \left(\frac{\textit{photosetter resolution}}{\textit{raster width}}\right)^2$$

Example: $\textit{Number of grey levels} = \left(\frac{2540}{150}\right)^2 = 16^2 = 286.7^{*)}$

*) The number of grey levels or colour levels per colour is limited to 256, as PostScript and all image processing packages cannot process more than 256 grey levels (8 bits).

32 grey levels

32 grey levels

64 grey levels

256 grey levels

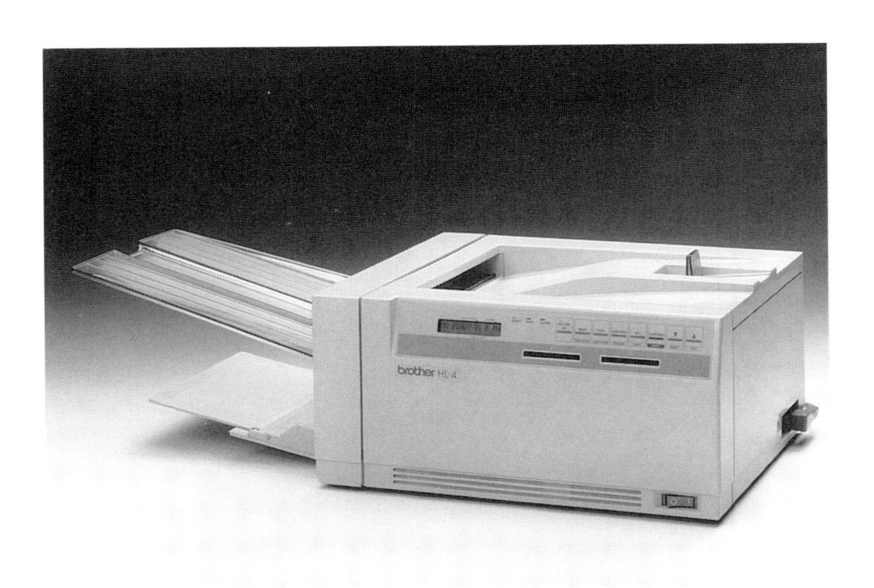

256 grey levels

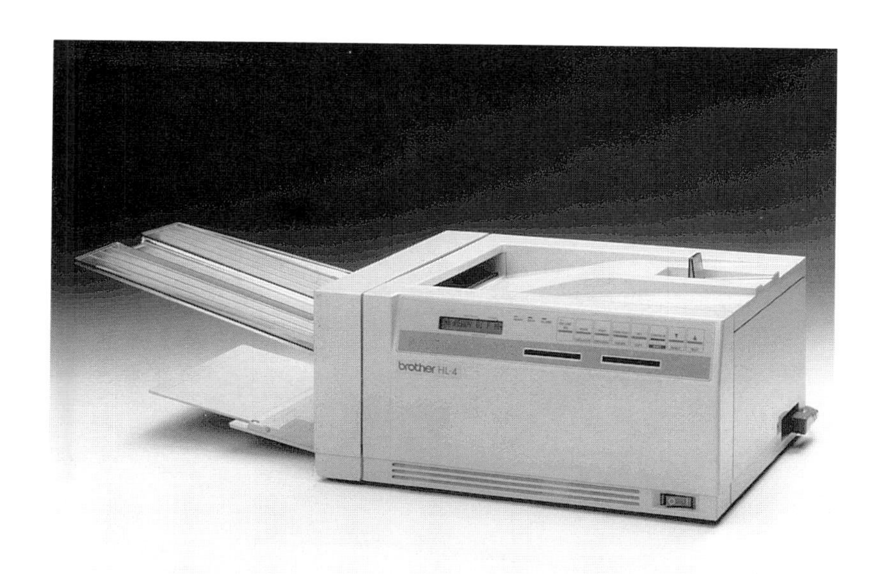

64 grey levels

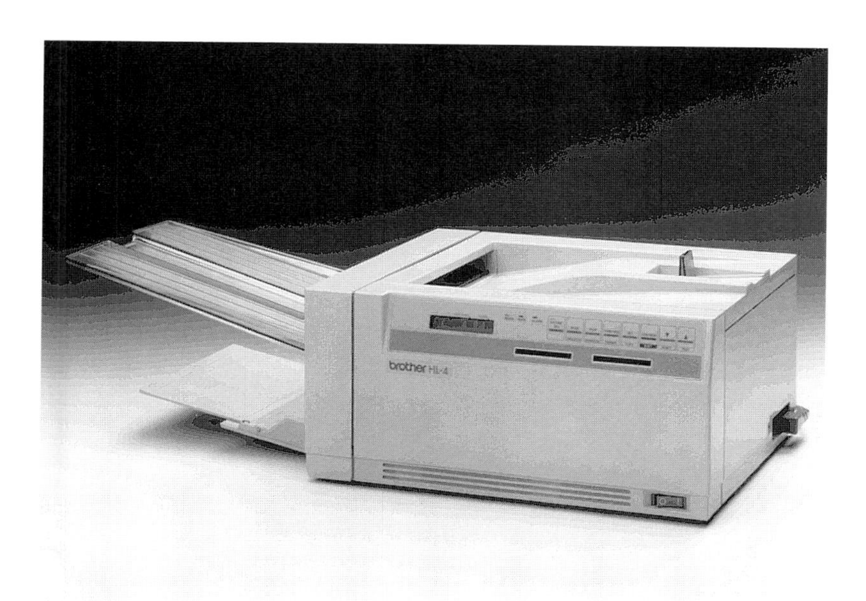

32 grey levels

Rasterising/dithering

As already explained, scanned grey dots need to be converted into a dot raster made up of dots of varying sizes. The creation of the raster dots from the scanned grey levels is known as 'rasterising' or 'dithering'. If rasterising is to take place, the raster width (distance between raster dots) used must be specified. Accordingly the dot matrix is calculated from the scanned raw data from the scanner. For this purpose various algorithms exist, but will not be discussed in more detail here. In the context here only the following is important.

When does rasterising take place?

There are various ways of rasterising. It is possible to rasterise during actual scanning or, on the other hand, to save the data first without rasterising and then to carry out the rasterisation separately at a later stage. Hence, before scanning, it must be specified whether the scanned data is to be saved rasterised (dithered) or not rasterised:

- Rasterising during scanning may result in loss in quality.

  If rasterising takes place during scanning then the desired raster and, above all, the eventual image size both need to be fixed. Therefore, the size of the image must be specified before scanning and must not be changed later. Creating the exact dot raster during scanning means that a later change, even only a small one, in the size of the image causes existing raster dots to move. In every case this will lead to a loss in quality (e.g. moiré patterns).

- Not rasterising during scanning implies 'freedom' and quality.

  If the scanned imaged is stored without rasterising (saving the real grey levels) it can be further processed later and reduced or enlarged without problem. During scanning it is simply the approximate final image size which needs to be known in order not to waste memory and to avoid long image calculation time when photosetting. Because the dot raster has not yet been created, the raster width may be determined at a later time using a different program. Rasterising pictures only takes place once the document has been completely laid out using a DTP package and is ready to be printed.

Throughout our explanations on the subject of image processing we assume that rasterising does not take place during scanning.

 *Never rasterise during scanning, but always save the original grey levels.*

Who rasterises?

As you must not rasterise during scanning, but only rasterised pictures can be printed, the question arises: who carries out the rasterising and when does it take place? Essentially it is true that rasterising must be the last step in the picture processing chain since it is primarily determined by the resolution of the output device (laser printer, photosetter):

- If a non-PostScript printer is connected, the actual printing program carries out rasterising automatically.
- If a PostScript printer is selected as the output device, the grey levels are transferred to the printer, and the printer itself carries out the rasterising.

  The PostScript printer, as well as the photosetter, incorporates a so-called RIP (Raster Image Processor) which carries out the rasterising of picture data (grey levels). This RIP is either built into the printing device or else connected in front as a separate unit (e.g. in the case of a photosetter). If the data is stored without having been rasterised, the same picture data may be used for printing a sample on a laser printer as well as for producing high-quality output from a photosetter. The photosetter then rasterises at 150 lpi (for example) and the laser printer at 53 lpi. The angle of the raster must always be specified at 45° for black and white pictures.
- An exception to this is the sublimation printer, which can print actual grey or colour levels and hence rasterisation is not necessary. Of course, you may also send rasterised picture data to such a printer, but you will obtain a bad printing result.

Scanner data format: TIFF must be used

Above, the advantages were described which are gained when you scan without rasterising and thereby store grey levels. The scanned data must, however, always be stored in the correct format, that is the TIFF format. Beware though, because data may also be stored rasterised using the TIFF format. Therefore, you must always ensure that the 'rasterise' or 'dither' function is not enabled in the scanner software. The conversion of TIFF files without loss in quality is possible, for example in a PostScript file (EPS file). TIFF (Tag Image File Format) has become established as the standard for storing pixel pictures. Many formats exist which are capable of storing black and white, greyscale or colour pictures. In order to stop this chaos Aldus, Microsoft and other leading scanner manufacturers agreed in 1986 on a single format for scanned pictures, that is TIFF. This format is used not only in the PC field, but also on Macintosh, Atari and Unix equipment.

Setting the raster

In order to calculate the required scanning resolution, the raster width and therefore also the output device must be known. For example, the current presetting of the output device may be used:

- 300 dpi laser printers usually have a preset raster width of 53 lpi (21 L/cm)
- Photosetters or RIPs (Raster Image Processors) from the firm Linotype (e.g. Linotronic 300) usually have a
  - preset raster width of 100 lpi (39 L/cm) at a resolution of 1270 dpi and a
  - preset raster width of 150 lpi (60 L/cm) at a resolution of 2540 dpi.
- Photosetters or RIPs from the firm AGFA-Compugraphic (CG 9400 PS) usually have a preset raster width of 100 lpi (39 L/cm) at resolutions of 1200 dpi and 2400 dpi.

⇨ ***Warning:*** *The values given for preset raster widths may be changed by the photosetting firm.*

In general the raster width which is entered into the DTP package is dependent on the quality of the paper to be used, so that the preset raster width does not necessarily have to be known. Some programs only allow the raster width to be set for pictures and not for grey areas. If these grey areas are to be photoset with a particular raster width, a suitable command for the desired raster width needs to be included later in the PostScript printing file. Otherwise the photosetting studio must be asked to change the preset raster width.

The preset raster width is hence only valid for the case where the raster width cannot be entered in the DTP software. For example, in Ventura Publisher the raster width may be entered under menu point 'Frame/image settings'. The raster angle for black and white greyscale pictures must always be 45°. The raster angle must always be entered. If only the raster angle is entered with the raster width left at 0, then it is usually possible to photoset or print using the current preset raster width of the output device.

Which raster width should be chosen?

Paper quality, printing process, raster width and photosetter resolution must all be matched to one another. The finer the raster, the higher the photosetter resolution and the higher the quality of the paper needs to be. A raster width of 150 lpi (60 L/cm) is standard for high-quality pictures using offset printing. The following

table gives only a preliminary idea. The problems involving 'raster width, photosetter resolution, paper quality and printing processes' will be described individually in detail later.

	Raster widths			
	53 lpi *) (21 L/cm)	75 lpi (29 L/cm)	100 lpi (39 L/cm)	150 lpi (60 L/cm)
Resolution of the output device	300 dpi (laser printer)	635 dpi (photosetter)	1270 dpi	2540 dpi (photosetter)
Camera-ready copy	Paper (or if necessary special film)	Photographic paper (film)	Film (photographic paper)	Film (photographic paper)
Printing process	Copying, Offset printing	Copying, Offset printing	Offset printing	Offset printing
Paper quality	Any paper	Any paper	Any paper	Coated paper
Application	Quick tasks, monitoring printouts, advertisements in daily newspapers (without photographs), small advertising brochures, etc.	High quality text, grey scale pictures of newspaper quality at most	Very high-quality text, average greyscale picture quality	High-quality print, advertisements, brochures, etc.

*) *lpi = lines per inch; L/cm lines per centimetre*

The raster counter

It may be desirable to measure the raster width of a printed picture, for example if you want to check the preset raster width of a photosetter or if you need to estimate the quality of a picture from a brochure that you wish to scan. Using a raster counter it is possible to measure the raster width of a photoset or printed picture easily.

If two rasters are placed one on top of the other, moiré patterns are produced. At those places where the raster widths of these two rasters exactly coincide, the moiré pattern disappears. A raster counter is therefore simply a clear plastic film on which

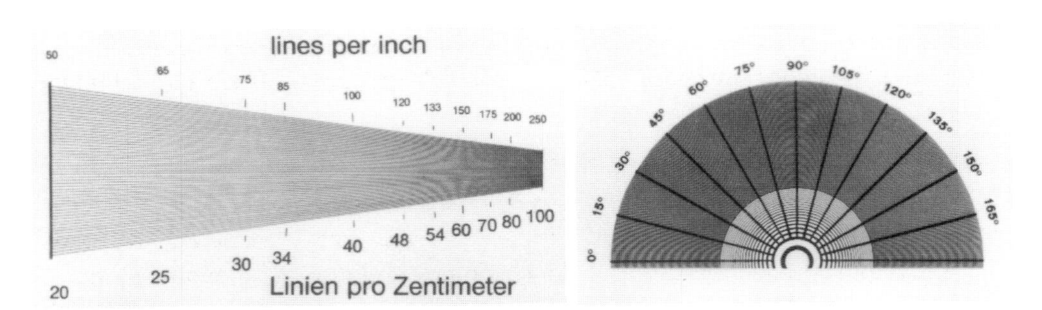

Owing to the propagation of the moiré pattern it is possible to measure the current raster width (left) and angle of the raster (right) of an original picture.

lines are printed at different fixed distances from each other. In practice these lines represent a line raster in which the raster width continually changes. If the raster counter (with the line raster) is placed on a picture which has already been rasterised, moiré patterns are produced.

Owing to the propagation of the moiré pattern it is now possible to determine the raster width of the original picture. The point at which the moiré lines are created is where the raster width is read. As the raster of the original picture is set at a specific angle, the raster counter must also be set at this angle. To achieve this the raster counter is simply turned until the moiré pattern is visible at its optimum. In single-colour originals the raster counter is rotated by 45°. In colour prints four rasters are placed one on top of the other, so that a raster width may be measured separately for each colour. For example, the raster counter may be set to 15° for blue, 75° for red, 0° for yellow and 45° for black.

Anyone who is concerned with the printing of photographs should work with a raster counter at least once, since only after you have observed the creation of moiré patterns with the help of a raster counter will you be able to understand the theory properly.

Difference between scanning and image resolution

If the printing process and the paper quality and hence the width of the raster are fixed, then the scanning resolution may be calculated. However, we must first introduce a new term, namely image resolution. A crucial factor affecting the end

result for the output (printer, photosetter) is not the scanning resolution, but the resolution of the image data set. The resolution of the picture file (TIFF file) is known as the image resolution.

In order to create a file with a particular image resolution, it is possible to scan at this desired resolution. Hence in most cases the image resolution will be identical to the scanning resolution. If the image resolution is subsequently calculated, the same value may appropriately be used for the scanning resolution. The two resolutions, however, must be differentiated when, for example, scanning takes place at a high resolution and subsequently the resolution is reduced in an image processing package. For this reason, it is important to know not only at which resolution scanning is taking place, but also which resolution the data set has. We will consider later in this book those cases where it is necessary to scan at a high resolution and subsequently reduce the resolution.

Optimisation factor for the calculation of image resolution

For the image and scanning resolution the specification of the raster width should suffice. If, for example, a picture is to be printed with a raster width of 150 lpi, an image or scanning resolution of 150 dpi should suffice.

PostScript interpreters, as well as many other printer drivers, use a particular calculation method to optimise pictures when rasterising. In this process, a grey level is not simply converted to a raster dot, but the neighbouring grey levels are taken into account in the calculation of raster dots. From this optimisation process the correction factor of the square root of 2, i.e. 1.4, is produced.

Since the memory requirement increases considerably when using the correction factor of 1.4, it is of course possible to decide whether you want to work with a lower correction factor. Experiments have shown that no serious loss in quality occurs even when a correction factor of 1.2 is used. Therefore if memory and photosetting costs need to be saved and if the desired image size in the DTP document is known precisely, then the factor of 1.2 may be used instead of 1.4. If the exact image size is not fixed before scanning, the image resolution (and hence the scanning resolution) may be selected to be higher than that calculated. The consequence of this, however, is that you will soon need a new larger hard disk and the photosetting costs will increase considerably.

Calculation of the image or scanning resolution

It will now be explained how the scanning process is carried out. First, the so-called pre-scan is carried out, i.e. the complete page is scanned at a low resolution. Next, the desired section is fixed by a frame. By means of this frame the scanner knows which area it should scan and what size the original is. Now what needs to be considered is how large the picture is going to be in the DTP document.

The eventual desired size of the picture in the DTP document may be taken into account by the scanner in one of two ways:

- Either the scanning resolution is changed correspondingly
- Or the size of the picture is not considered when calculating the scanning resolution and hence the desired image size needs to be entered separately into the software.

In the end it does not matter whether the scanning resolution is changed with respect to the picture size or whether the picture size is entered separately. In both cases, the scanning resolution is affected. You may check whether this is true for your particular scanner by carrying out the following. First scan at 75 dpi with an image size of 200% and then scan again at 150 dpi with an image size of 100%. In one case the image size is doubled and in the other the scanning resolution is doubled. Therefore, the same scanning resolution is used in both cases and the resultant image files will be the same size.

The resolution used for scanning should not be higher than the physical resolution offered by the scanner, except where the scanner is able to interpolate grey levels. A higher resolution may then also be selected. Usually, however, interpolation of the grey levels is only possible using an image processing package, so we will come back to this function later. If possible, it is better to use a larger original than to select too high a resolution. However, bear in mind that all the specifications in this book are only valid for scanned photographs saved without rasterising (without dithering)!

Below, two possibilities are described on how the desired image size in the DTP document may be taken into account when calculating the image resolution.

a) Resolution only specified

It is possible to leave the image size set at 100%. Furthermore, scanner software exists in which the desired image size cannot in fact be entered. In this case the desired image size for the DTP document must be taken into account when calculating the image resolution. If the image is to be larger than the original, then the resolution is increased accordingly. If the image is to be smaller than the original, the resolution is decreased. The scanning resolution to be set must be identical to the required image resolution and may be calculated as follows:

$$Image\ resolution\ (dpi) = raster\ width\ (lpi) \times 1.4 \times \frac{desired\ size}{size\ of\ original}$$

Here the image resolution (= scanning resolution) is dependent on the size of the image. If the desired image size is fixed precisely beforehand and if memory needs to be saved, then the factor of 1.2 may be used instead of 1.4. For determining the size of the picture, the length of one side only is used.

Example:

$$Image\ resolution = 150\ lpi \times 1.4 \times \frac{8\ cm}{14\ cm} = 120\ dpi$$

$$Scanning\ resolution = image\ resolution = 120\ dpi$$

b) Image size and resolution specified

It is simpler if it is possible to specify the image size in addition to the scanning resolution. If the image size is entered in the scanner software then the scanning resolution is automatically changed without user intervention. If a larger value than 100% is entered, then the resolution is increased. If a lower value than 100% is entered, the scanning resolution is reduced accordingly. Suppose a scanning resolution of 200 dpi is required and a desired image size of 150% is entered; then in reality scanning takes place at a resolution of 150% of 200 dpi = 300 dpi.

The scanning resolution to be set is calculated as follows:

$$Image\ resolution = raster\ width \times 1.4\ (or\ 1.2)$$

Example:

$$Image\ resolution = 150\ lpi \times 1.4 = 210\ dpi$$

$$Scanning\ resolution = image\ resolution = 120\ dpi$$

Using the formula above the image size must be entered subsequently. Various ways now exist for taking the eventual desired image size into account. If the desired image size can only be entered as a percentage, then a suitable value can be calculated as follows:

$$Image\ size\ in\ \% = \frac{desired\ size}{size\ of\ original} \times 100$$

Example: A picture of size 14 cm × 11 cm is to be scanned. The eventual image size should be approximately 8 cm × 6 cm. In order to calculate the image size as a percentage, a length of one side of the picture is used:

$$Image\ size\ in\ \% = \frac{8\ cm}{14\ cm} \times 100 = 57\%$$

Of course it is ideal if the scanner software allows you to enter the image size directly in centimetres. In this case the desired image size is not entered as a percentage, but simply as the length of a side of the eventual image. The software then calculates the second length accordingly so that the picture is not distorted.

The scanning resolution to be entered depends on the raster width and hence in general may remain the same. Therefore, all that remains is to enter the desired image size in centimetres and then scanning can take place. When you use this method (specifying the image resolution and image size) you do not of course know at which resolution scanning really takes place. This needs to be calculated separately. If a picture is to be enlarged, you should check that the set scanning resolution is not too high for the scanner. The set scanning resolution should not in general be any higher than the physical resolution of the scanner. Only for a set image size of 100% will scanning also take place at the set resolution. Hence, if a picture is to be greatly enlarged, it is sensible to calculate the real (effective) scanning resolution as follows:

$$\textit{Effektive scanning resolution} = \textit{set resolution} \times \frac{\textit{set image size in } \%}{100}$$

Memory requirement

A note on the amount of memory required: if the resolution of the scanner is doubled, the memory requirement is quadrupled. There is a simple explanation for this. The resolution is always specified for one axis. If, for example, scanning takes place at a resolution of 100 dpi, an image is produced with a resolution of 100 dpi (image width) × 100 dpi (image height). This corresponds to a resolution of 10 000 dots per square inch.

If the resolution is now doubled to 200 dpi, this results in a resolution of 200 dpi × 200 dpi = 40 000 dots per square inch. With 40 000 dots you have four times as many dots, even though the resolution was only doubled. This is also the reason why even a small increase in the scanning resolution increases the size of the image data set enormously.

⇨ *As memory is squared when it is increased, it is recommended that the image size is calculated as precisely as possible.*

Chapter 7
Scanning rasterised pictures

How are moiré patterns created?

If you have read the book up to this point, you now know what rasterised images are. If a photograph is to be printed, it must be converted to a dot raster. It is often overlooked that pictures from postcards, brochures, magazines, etc., have already been rasterised. In high-quality prints, however, the raster is so fine that you can only detect it by using a magnifying glass. Suppose you wish to place a picture of a product in an advertising leaflet and as an original you only have the brochure from the manufacturer. If you scan the picture from the brochure, then undesirable moiré patterns are produced.

What are moiré patterns? They are undesirable regular patterns (interference patterns) which are produced when, instead of grey levels, a regular fine pattern is scanned. A picture from a magazine consists not of grey levels, but of a pattern of black dots of varying sizes and so such an interference pattern can hardly be avoided. If a scanner scans a raster image, the scanner sensor will on one occasion record a raster dot at its centre and on another the space between two raster dots. If the scanner selects the raster dot closer to the centre the level is darker but if it selects the dot closer to the edge a white area will be detected and the level will be lighter. Hence the same level of brightness is once selected once as too dark and once as too light. Depending on the raster width, scanning resolution and change in image size various patterns are produced.

Hence if a picture which has already been rasterised is scanned, moiré patterns occur. However, with suitable knowledge, even very good results may still be achieved here. Expensive professional scanners or DTP camera scanners allow you to adjust the optics so that the individual dots are no longer so clearly recognised. It is not possible, however, to adjust the optics in DTP scanners. We may achieve this effect by using an image processing software.

Scanned from a four-colour brochure at 100 dpi: moiré patterns and fuzzy lines were the result.

Scanned at 300 dpi with the resolution later reduced to 100 dpi: moiré patterns can no longer be seen. The irregularities in the lines and edges have also been removed.

(Source: brochure on photosetters and RIPs from AGFA Compugraphic GmbH)

Top picture: Scanned from a four-colour brochure at 140 dpi.
Moiré patterns are produced.
Bottom picture: Scanned at 300 dpi with the resolution later reduced to 140 dpi.

Scanning rasterised originals

First of all it must be stated that the usual calculations for scanning resolution do not normally apply to the scanning of rasterised pictures. If a raster original is scanned, one raster dot is selected nearer its centre and the next nearer its edge. Therefore, one scanned grey level is too dark and the next too light. From both these values it is possible to calculate a grey level which corresponds closely to the brightness level of the original. The picture becomes less sharp in doing this. As already mentioned, scanners exist which achieve this 'unsharpening' by means of the hardware (optics), whilst we will achieve it through the software.

Let us now consider how a scanner scans a picture. If, for example, a scanning resolution of 100 dpi is set, then a 400 dpi scanner only scans using every fourth sensor. Therefore, parts of the picture are left which are not scanned. However, in order to achieve optimum picture quality, each part of the picture needs to be recorded to calculate the correct grey level from several grey levels. The physical resolution should therefore be used when scanning raster images in order to scan the complete area uniformly. Consequently, if you are using a 300 dpi scanner you should set a resolution of 300 dpi and an image size of 100%. In this way you will produce a data set which may be further processed to an optimum degree.

However, it is only sensible to scan at high resolution if, after the individual grey levels have been balanced, the data set can again be reduced in size. If it is not possible to reduce the resolution, then for reasons of cost you will need to refrain from scanning at high resolution.

Hence the most important rule when scanning raster images is:

The optical resolution of the scanner should be used for scanning rasterised originals if it is possible to reduce the size of the data set afterwards.

As the scanning of rasterised originals is often completely overlooked, or dealt with insufficiently by manufacturers of DTP hardware and software, a suitable image processing packaged is required which offers the facility for reducing the data. Scanning at high resolution and the subsequent data reduction are only necessary because the scanning procedure in many scanners is not optimised for dealing with low resolutions. At lower scanning resolutions, some of the CCD sensors are often simply switched off and the picture is then scanned with gaps. Only scanners using HRR technology (High Resolution Reduction) ensure that a picture of optimum quality is produced even at low resolutions.

Reduction of data with interpolation

If scanning has taken place at high resolution the size of the data set needs to be reduced to that which is actually required. However, this cannot be achieved by a simple reduction in the amount of data, as otherwise you would have been able initially to scan with a lower resolution. When reducing data, the function 'interpolation' or 'SmartSizing' must be enabled.

The reduction of data with active interpolation gives rise to the following. Before the resolution is reduced and a grey level removed, a new calculation of the remaining grey levels is carried out. Through this reduction it is possible to achieve a considerable improvement in quality. As a result of the interpolation the moiré patterns are removed, since a nearly correct grey level may be calculated from a grey level which is too dark and one which is too light.

Neighbouring grey levels with stark differences in brightness are thus brought closer together. For this reason pictures must not be sharpened, otherwise the differences in brightness are amplified again. Moiré patterns are reduced when scanning is carried out at the optical resolution and afterwards a data reduction with active interpolation may be performed.

⇨ *In general the picture must not be sharpened.*

⇨ *Moiré patterns are not only removed by means of image processing, but also reduced a little by setting a coarser raster (e.g. instead of 150 lpi 'only' 120 lpi) in the DTP software.*

Hence an image processing package is required which reduces the scanning resolution of 300 or 400 dpi to the required resolution. In order to calculate the required image resolution two methods exist, as already mentioned. Either the image resolution only is changed or the image resolution and the image size are changed. Which method is used depends on the capabilities of the software. The simplest method is to enter both the image resolution and the desired image size, in centimetres or as a percentage. The required image resolution is calculated as before:

$$\textit{Image resolution} = \textit{raster width} \times 1.4 \ (\textit{or}\ 1.2)$$

The image size must be entered separately in the image processing package. If it is not possible to enter the image size in the image processing package then, at the time the image resolution is determined, the image size desired for the DTP document must also be taken into account:

$$Image\ resolution = raster\ width \times 1.4 \times \frac{desired\ size}{size\ of\ original}$$

The image resolution is dependent on the image size here. Instead of the factor 1.4, the factor 1.2 may also be used in the calculation, resulting in a slight loss in quality. The calculated values may be entered into the image processing package. The data set is then reduced in correspondence with these values. A reduction by half in the resolution reduces the memory requirement by a factor of four. It is therefore worth while to determine beforehand the exact image size desired.

Moiré patterns as well as the result of data reduction may be examined roughly on the monitor. However, a problem with this is that the moiré effects depend on the resolution of the output device and so different patterns appear on the monitor than are present later on the print. If the picture is reduced more and more, various different patterns appear depending on the size of the picture. If the picture is then photoset, different patterns appear again, since the photosetter has a resolution different to that of the screen.

The best way of assessing a picture is when it is shown enlarged on the screen. It is then possible to see the sharp differences in brightness between neighbouring grey levels. If an interpolation is carried out, the distracting diamonds blur more and more. Even if the interference patterns are still slightly recognisable on the enlarged image, they are hardly visible once the image has been printed.

Data reduction and smoothing

In the following section we include a few more notes on what to do if your image processing program is not capable of reducing data by interpolation or if the moiré patterns have to be decreased even more. The creation of moiré patterns is affected by the quality of the original as well as the size of the image. The finer the raster on the original, the better the result will be. Even if the original is quite large and the eventual image size is smaller, a better-quality picture may be attained by means of data reduction. In the most unfavourable case it may be necessary, in spite of an interpolation function being present, to use additional smoothing.

If your program does not possess an interpolation function, then the image must be processed before the data is reduced. By scanning rasterised originals sharp differences are produced in the brightness between neighbouring grey levels, which cause moiré patterns to be created. These sharp differences in brightness may also be reduced if the image is processed using a smoothing filter.

Various types of filter usually exist: 'normal smoothing', 'strong smoothing' and 'extreme smoothing'. Which filter and how often it should be used is dependent both on the quality of the original and how great the difference is between the set scanning resolution and the required image resolution. However, be economical in your use of filters because, if filtering is too severe, pictures will become too unsharp. It is recommended that you work with a normal smoothing filter, examine the result on the monitor and then, if necessary, carry out the smoothing again. Minor moiré effects disappear on photosetting, since the photosetter also interpolates as it rasterises. After smoothing the contents of the picture have been changed and it is less sharp and softer, but the data set has, as beforehand and unnecessarily so, the same size as directly after scanning. Therefore the data set needs to be reduced to the size that is actually required. However, if you have plenty of time and money, or if you do not have a sensible image processing program, you may omit the data reduction and allow the photosetter to carry out the interpolation. Furthermore, this may be the reason why many users work with scanning resolutions which are too high. Often the task of improving the quality of the picture is left to the output device, with the disadvantage that very large quantities of data need to be processed.

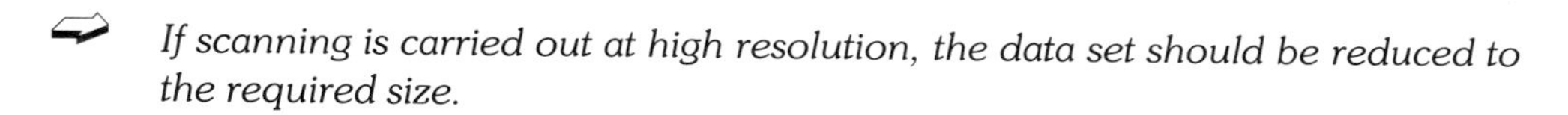

If scanning is carried out at high resolution, the data set should be reduced to the required size.

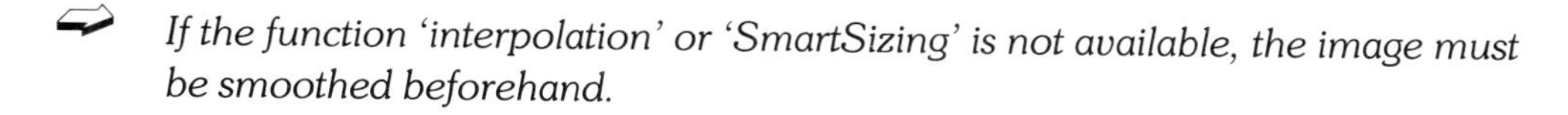

If the function 'interpolation' or 'SmartSizing' is not available, the image must be smoothed beforehand.

The following pictures demonstrate moiré problems. However, note that moiré patterns are always different, as the results depend on the quality of the original, scanning resolution, size of the original, size of the picture in the DTP document and resolution of the output device (photosetter).

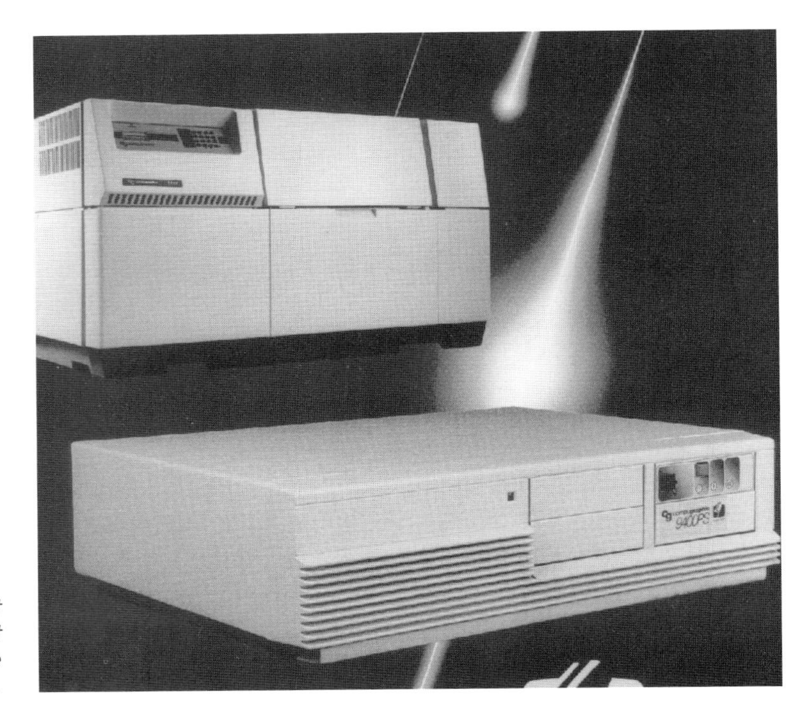

Scanned at 110 dpi without further image processing.

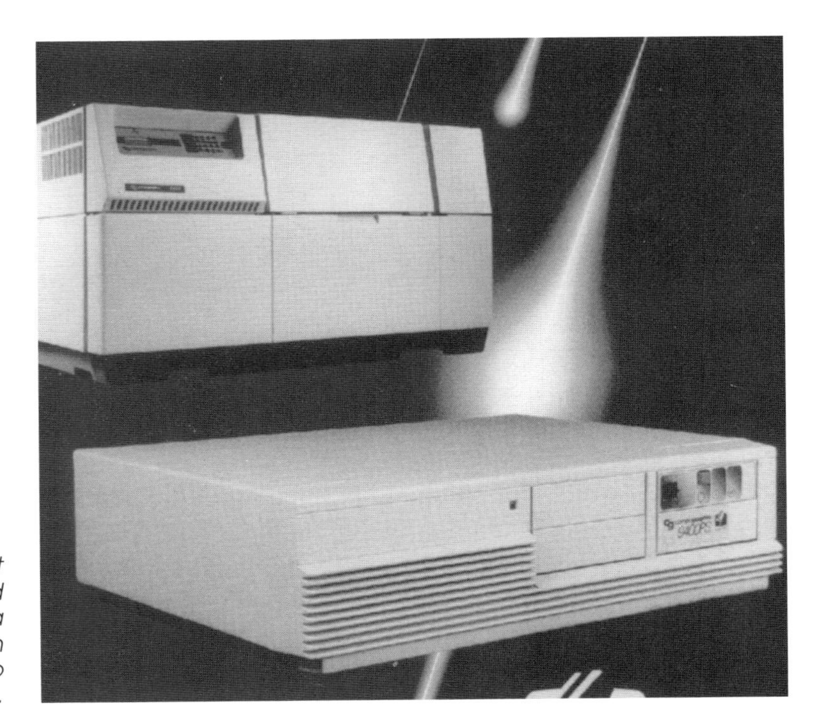

Scanned at 300 dpi and then data reduced (with interpolation) to 110 dpi.

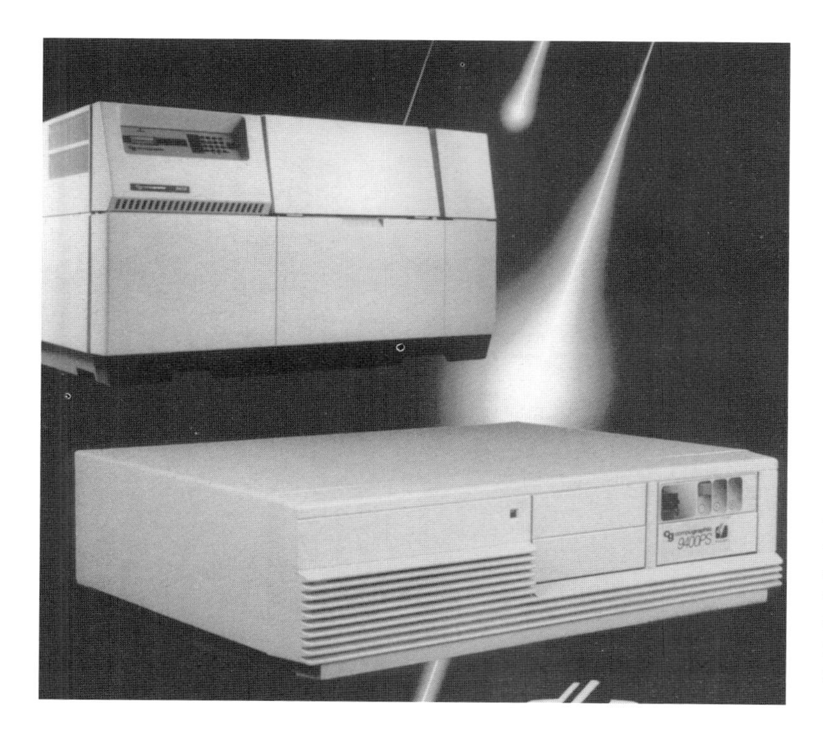

Scanned at 300 dpi, smoothed twice and then data reduced (without interpolation) to 110 dpi.

Chapter 8

Camera-ready copy and photosetting

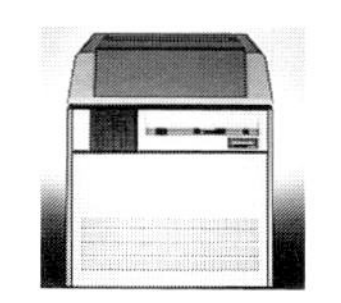

Which camera-ready copy?

The paper or film copy needed by the printer for reproduction is known as the camera-ready copy. By using a DTP system camera-ready copy may be created simply and cheaply. The reproduction of camera-ready copy which contains only text and drawings is no problem and a book does not need to be written on the subject, even though much can surely be said regarding the layout and quality of text. It is, however, always a problem when photographs are to be reproduced. As simple as it is to read the pictures into a computer, it is considerably more difficult to print them. As they need to be rasterised in order to be reproduced, either a printer with a very high resolution (from about 1000 dpi) or a printer which is capable of varying the sizes of dots (at least 64 different sizes of dots) is needed.

Camera-ready copy with text and pictures of optimum quality can at present be obtained only by using a photosetter. As these photosetters are expensive, photosetting studios offer suitable services. The data is sent on a disk to the photosetting studio and photoset film is returned. However, even with relatively inexpensive printers good-quality pictures may be obtained. For example, there are ink jet printers which are capable of changing the size of dots and so-called sublimation printers which can print real grey or colour levels. However, these printers are not suitable for the creation of camera-ready copy for reproduction, as copy used for printing may only be black and white, rasterised without grey levels.

The following may be used as camera-ready copy:

❒ Prints from a laser printer

If normal paper is used for printing, the printing firm needs to produce a printing plate or film. However, special paper/printing film does exist for the laser printer which may be used directly for small offset printing. The production of a printing plate is then not required.

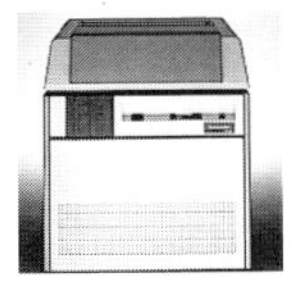

- ❐ Photosetting of data on to paper or film

  For professional printing results, the data is photoset at high resolution on to film or photographic paper. As these photosetters are very expensive, a large number of organisations offer photosetting services at a reasonable price. From the film the printing firm then produces a plate for printing.

- ❐ Photosetting of data on to direct printing film

  If the data is photoset onto so-called direct printing film, the production of a printing plate is no longer necessary. Costs may then be reduced. However, if a printing film is used a print run of only up to 10 000 pages may be carried out. The raster width may not exceed a maximum of 120 lpi (47 L/cm).

Camera-ready copy from your own printer

Even prints on paper can be used as camera-ready copy. However, the printing firm still has to produce a printing plate or film, using a process or plate camera, directly from the paper copy. To produce a printing plate or film various processes exist which are more or less time and cost consuming. It depends on how well the printing firm is equipped.

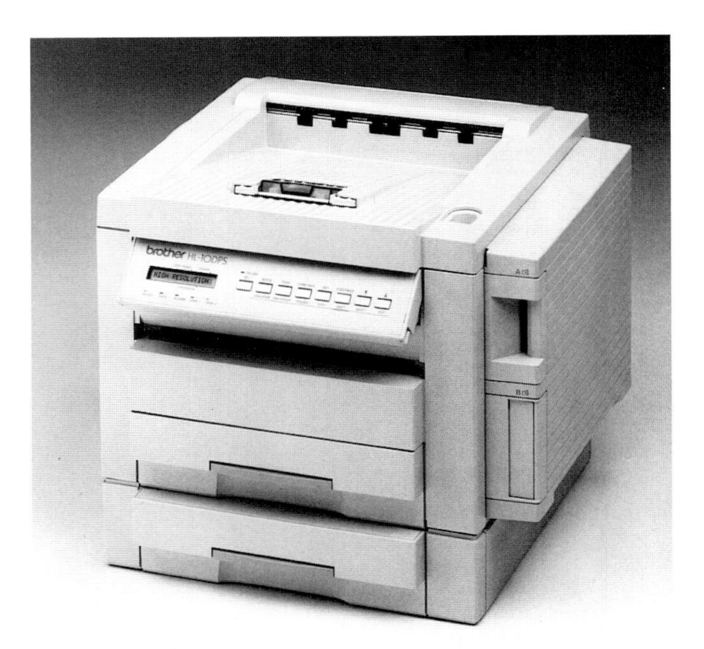

PostScript laser printer

Photograph: Brother

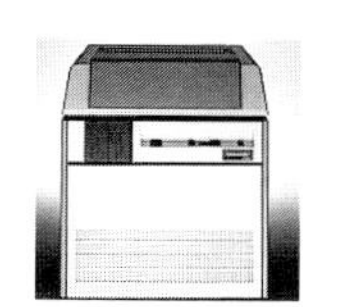

However, special paper/printing film also exists, for use in the laser printer instead of normal paper, which may then be used directly for reproduction. The production of printing plates or film by the printing firm is then not necessary. Even the offset copy prints offered by photocopying shops are produced using normal offset printing machines. Hence here also a printing film is produced from the paper copy.

If people were once amazed at the quality of the laser printer, nowadays the 'low' resolution of 300 dpi is often not sufficient to meet requirements. Higher quality is expected from brochures and large-scale advertising campaigns, so that in many cases it is necessary to obtain professional camera-ready copy by means of photosetting the data. However, cases do exist where photosetting is not necessarily appropriate, yet in spite of this the best quality possible is still desired. For this purpose laser printers exist which operate at higher resolutions or vary the size of pixels. The improved quality in text due to changing the pixel size may be explained simply. The laser beam is affected in both position and size in such a way that gaps and zigzags on diagonal lines are almost completely compensated for. Curves become rounder and the transitions smoother. With high resolution laser printers (1200 dpi high quality text, drawings and greyscale levels may be obtained. For example, with a DMV laser printer greyscale pictures can be printed with a raster width of 150 lpi and 256 grey levels (which corresponds to 2400 dpi). However, these prints are less suitable as camera-ready copy when they contain greyscale pictures. Laser printer output can only be used as camera-ready copy when the raster width of the photographs is maximum of 100 lpi.

For the rasterised output of greyscale pictures, special ink jet printers exist which can create 64 different dot sizes and hence 64 levels, even in colour. Relatively good pictures may be obtained with these printers. However, the disadvantage of an ink jet printer is that its resolution of about 160 dpi is relatively low for the representation of text. These printouts are not (yet) suitable as camera-ready copy for reproduction. Furthermore so-called video printers or sublimation printers are used for printing greyscale pictures. Since they output real grey levels, these printers can even produce colour pictures of high quality. However, these printouts cannot be used as camera-ready copy, as rasterised pictures are required for printing.

How is professional camera-ready copy produced?

In order to produce professional camera-ready copy the data needs to be photoset. Normally the data is photoset on to film. The printing firm then produces a printing plate from this film. Instead of photosetting the data on to film it may be photoset on to direct printing film. If direct printing film is used printing may be carried out

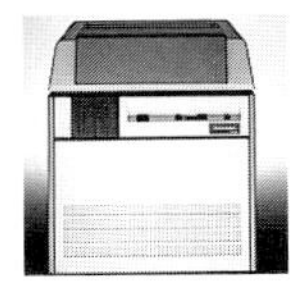

immediately, so that the production of a printing plate is omitted. Direct printing film does, however, have disadvantages with respect to quality, print run size and further processing. More will be said about this in Chapter 9. Here, we are primarily concerned with the creation of camera-ready copy (film). How is the page layout photoset? In principle it is very simple. You send your data to a photosetting service and receive back the paper or film. However, several issues need to be taken into account as explained in the following.

Use of PostScript printing data

Even if you do not possess a PostScript printer you can still have your data photoset. In the DTP software you need to specify a PostScript printer. All the PostScript options are now available. You will no longer be able to print on a non-PostScript printer, but all the line and page breaks will be shown on the screen exactly as they will appear once the data is photoset. In some DTP programs the printer type may be selected independently of the so-called letter spacing table. This table holds information on the text used, such as the distance between letters. If a PostScript

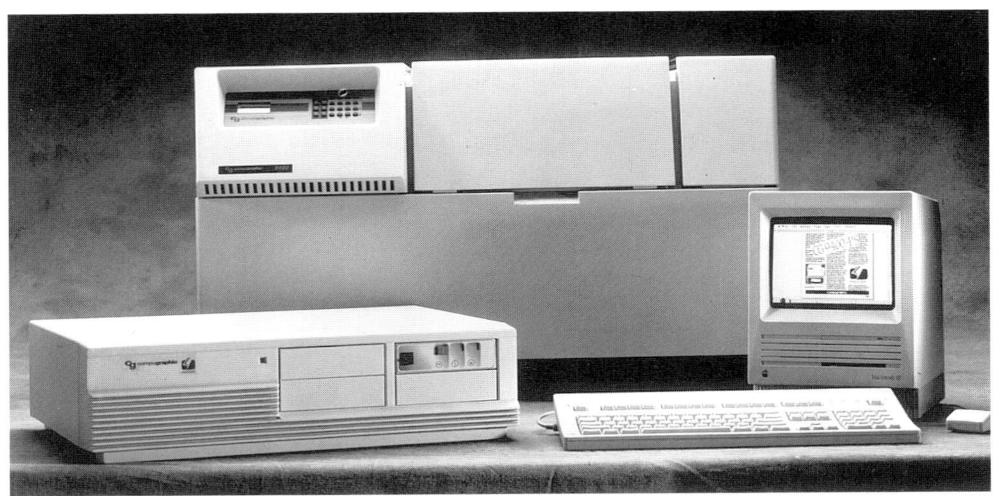

Photograph: AGFA-Compugraphic GmbH

Front left: RIP (Raster Image Processor), a computer which produces a pixel file from the PostScript file for the photosetter. Centre: photosetter, which blackens film or photographic paper by means of a laser beam. Right: personal computer, which sends the PostScript file to the RIP.
The photosetting studio must connect various types of computer (e.g. MS-DOS personal computers, Apple Macintoshes, etc.) to the RIP in order to send current data to the RIP. In addition, the disks that current DTP customers use must be built into the personal computer, for example 5.25" and 3.5" disks, 20 Mbyte disks, 40 Mbyte streamers, etc.

letter spacing table and a non-PostScript printer are installed, then the PostScript data may also be printed by a non-PostScript printer. As a result page and line breaks occur as they would on a PostScript printer. However, a correct printout of the font and text size is not possible.

If you want to make sure that the line and page breaks are photoset exactly as you inserted them, then you need to send the PostScript printing file to the photosetter. Often, it is also possible to send the original DTP data to the photosetting studio, but the studio must then have the relevant software installed. As different program versions and fonts are supplied by different manufacturers errors and other effects may creep in very easily. In any case, many photosetting studios only accept PostScript printing files.

How do I create a PostScript printing file? In order to create a PostScript printing file you should print the desired pages as normal but, instead of sending them to the printer, request the software to redirect them to a file on the hard disk. This file may then be printed by any PostScript printer and photoset by any photosetting studio. Try it: send the PostScript file created by using the copy command to your PostScript printer.

How are the files sent?

As long as you only have text or a few pictures, data may be transferred readily using disks; 1.2 Mbyte capacity is normally sufficient. However, as soon as you have many pictures, 1.2 Mbytes is no longer sufficient. This hurdle may initially be bypassed

After photosetting, the film or photographic paper needs to be developed. The picture on the left shows a developing machine.
Photograph: Agfa

since, for example, a separate printing file is created for each page. It is in any case recommended that large blocks are always split up into smaller ones. Generally, it is more convenient to work with a backup program which on one hand compresses the data and which, on the other hand, allows a large file to be partitioned over several disks. On the whole, effective operation only takes place when using a data storage device which has at least 10 Mbytes of memory. Nowadays such systems are both convenient and inexpensive. A tape streamer may for example be connected to the disk drive controller to facilitate writing data to small portable cassettes with a capacity of 40 to 60 Mbytes.

It is even more convenient if, for example, disks of 20 Mbyte capacity can be used. Using high-capacity disks enables fast direct access to the data. With these 20 Mbyte drives it is even possible to read the usual (up to now) 3.5" 720 kbyte and 1.44 Mbyte disks. The optimum, widely used SyQuest or Bernoulli cartridge units can save 40 or 90 Mbytes. However, the best storage device is of little use if the photosetting studio does not use the same one. Therefore, before you buy a suitable drive, consult your photosetting studio.

Is the picture data correct?

If you would like to have a scanned photograph photoset it should be loaded into the DTP program as a TIFF file. Before the printing file is created, care must be taken to ensure that the picture data is properly set in your DTP software. The parameters for rasterising must be set as follows:

- Angle of raster

  An angle of 45° is to be set for black and white pictures (other values are used in colour pictures).

- Raster width

  The raster width entered in the DTP software must correspond to the paper and printing plate used. If no value is entered, the presetting of the output device is usually adequate. Which presetting is active in an individual device depends on whether the photosetting company alters the presettings of the RIP manufacturer.

Lines too thin?

Do not use any line thicknesses of less than 0.1 mm. If you select thinner lines, they will be photoset, but can no longer be printed. The printout from a laser printer is misleading, since all lines with a thickness of less than 0.1 mm are printed regardless with a thickness of 0.1 mm. The thinnest line for a photosetter is about 0.02 mm.

Photosetting

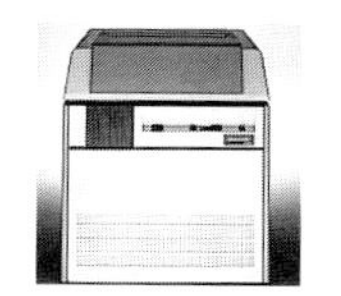

Anyone who has had data photoset for the first time soon appreciates the benefits of experience. Do not initially send huge documents to be photoset, but carry out several trials first.

Errors occurring in photosetting

Unfortunately, even photosetting is not without its problems. The fact that the scanning and printing of greyscale pictures is generally assumed to be difficult has, to date, resulted in relatively little camera-ready copy containing photographs being produced by means of DTP. After surveying photosetting companies we deduced that many attempts had been made, but that many people were disappointed with the results and soon gave up. Therefore, if you wish to attain pictures of outstanding quality, you need to choose a photosetting studio which understands the problems associated with photosetting greyscale pictures. With standard film the individual raster dots for a photograph become slightly larger since the edges of the dots are also partly photoset, i.e. the edges of individual raster dots become too blurred. The effect of this is that the complete picture becomes much darker and problems may also occur in producing the printing plate.

If the paper or film is too soft good results may still be obtained for line drawings and text, but it is a matter of chance as to whether good reproduction of photographs is achieved. If, for example, a 40% grey level is photoset onto a line film with an uncalibrated photosetter, then a grey level of more than 60% may be created.

For photosetting greyscale pictures suitable film and developer with a steep gradation are used. This produces a sharp raster dot and so errors occuring during photosetting will be negligible. In principle, a differentiation is made between line film (soft material for lines and characters), lith film (hard, steep material for lithographic applications such as producing raster images) which is not easy to handle, and hybrid film which has the characteristics of a lith film but is considerably easier to handle. Ideally, when greyscale pictures are photoset, the normal film available for the photosetting of text and drawings should not be used. However, as film material with lith characteristic is very expensive and not easy to handle, there are only a few photosetting studios which use this outstanding material.

Nevertheless, even with standard film very good results can be obtained since line film has been steadily improved over the past few years. In addition the dot sharpness of the photosetter has been improved and so, nowadays, it is the calibration of the photsetter and the developing process, and not the film, which is decisive. When mention is made of the calibration of the photosetter it is not software calibration which is implied. When calibrating via software the incorrect adjustment

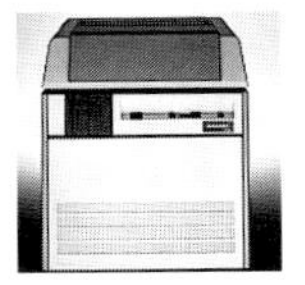

of the photosetter is unaffected, since only the data set is altered, namely until the desired grey tone is achieved. However, by changing the data set the number of grey levels is considerably reduced and this will naturally lead to poor-quality pictures. Therefore, we do not recommend calibration by software, but instead we recommend that the photosetter and the developing machine are adjusted correctly.

The light intensity of the photosetter must be adjusted and the complete developing process should be regularly calibrated using a densitometer. However, some photosetting studios only vaguely know what a densitometer is and consequently may also contribute towards producing poor-quality pictures. Therefore, when mention is made here of photosetting onto film with lith characteristic, a correctly adjusted photosetter with suitable film and developer is implied, whereby distortion of the picture will be avoided. This can be achieved by using line and lith films as well as hybrid films.

Trials

When photosetting tests are carried out care should be taken to ensure that the same material is used as required later for printing. The know-how and level of help obtainable from individual photosetting studios varies greatly, so it is not only the price of photosetting which needs to be considered. However, an optimum photosetting result does not guarantee that the picture will retain the same quality once it is printed. In fact, when printing is carried out another increase in half-tone occurs, so that the picture is made darker by the printing process. This must also be taken into account when actually processing the picture. See Chapter 9 for further information.

It is especially sensible to carry out several trials first if you wish to photoset scanned greyscale pictures, in order to check your scanner and photosetter settings. Do not make the pictures too large in the DTP software, only scan small sections of pictures and adhere to the image resolutions specified in this book. In this way the amount of memory required and the cost of your trials will not be too high. Also, create a sample file with various line thicknesses and grey areas and then have this photoset. Then you will have a sample from which it will be easier to gauge your settings. You will find that a page which has been photoset looks completely different to the corresponding printout from a laser printer.

As already mentioned briefly, pictures and grey areas become darker on photosetting and printing. If photosetting on paper, the grey areas become darker than if photosetting on film. However, the type of photographic paper as well as the settings of the photosetter and developer also influence the result. If you use paper for photosetting your tests and then use film with lith characteristic for your printing results, you will have simulated very roughly the printing process. You will therefore have an initial rough idea of how the results will appear after printing. However, you must not make the mistake of optimising your pictures on paper. We will return to these relationships in more detail later.

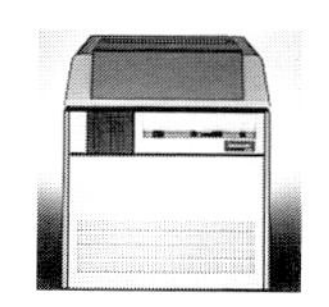

Differences in cost

We found considerable differences in price for photosetting, above all for files containing scanned pictures. It is therefore worth while obtaining several quotes. But comparing prices is complicated by the fact that some photosetting studios calculate costs on the duration of photosetting time used. For a photosetting studio using a slow RIP and PC, perhaps even sending the data from the PC to the RIP via a serial interface, you should not be surprised at high photosetting costs, even if the quote seemed reasonable initially. It is best if you agree fixed prices, or prices which are dependent on the size of the data file (PostScript), with the photosetting studio. You will then know beforehand exactly how much it is going to cost you.

Specifications for the photosetting studio

The photosetting studio requires several details so that the PostScript file can be correctly photoset. It is also sensible to send a printout from the laser printer to accompany the data.

You should specify how many pages each of your files contains. A further important specification is the desired photosetting resolution (e.g. Linotronic 300: 635 dpi, 1270 dpi or 2540 dpi). Equally you must specify whether photosetting is to be carried out to photographic paper or to film. If photosetting to film, the photosetting studio will need to know whether or not the photosetting is to take place with the pages reversed, positive or negative. For example, for the offset printing process the film is positively, pages reversed photoset. However, you should consult the printing firm beforehand to determine whether it works with positive or negative printing plates. In the USA and Scandinavian countries many printers work with negative printing plates, so they need to photoset the film negative with pages reversed.

However, it is possible to achieve higher-quality pictures with positive printing plates. Negative printing plates have only one advantage: films may be easily mounted to form the complete page at the light table, because the cut edges due to film mounting cannot be seen. If positive films are not managed carefully, the cut edges of the films may be visible. You will also be familiar with this problem from photocopying. If you form a page from several pieces of paper using glue, then the cut edges may be seen

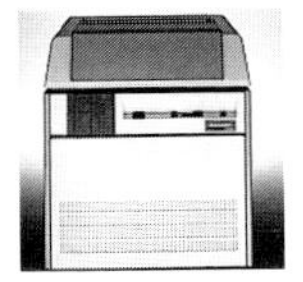

on copying. However, since we are photosetting a complete page with both text and pictures on to film, the cut edges do not exist and hence negative printing plates are of no advantage to us.

If you desire very high-quality greyscale pictures, you must expressly ensure that the photosetting studio is aware that the calibration of their equipment should not be carried out by means of software und that only minor distortion of the picture is permitted.

The photosetting result should fulfil the following requirements:

- The change in half-tone should be less than 5%.
- The density of the basic veil should be less than log 0.06.
- The full-tone density should be at least log 2.4.
- The film must be free of bends, scratches or marks which could affect the printing plate.

In Chapter 12 you will find a standard photosetting order form.

What the photosetting studio needs to take into account

In order to obtain optimum picture quality it is important that the correct film material is used. As already mentioned, film materials exist with line characteristic (films with flat gradation for texts and drawings) and with lith characteristic (special line films, hybrid films and lith films: with steep gradation for raster images).

However, it is not only the film material which is important, but also the developing chemicals, the correct setting of the photosetter and the developing process. The calibration of the photosetting process is important to achieve optimum quality. The photosetting studio must control the photosetting process and if necessary reset it after each change in paper or film, after each change of chemicals and after each service.

Initial testing of the developing machine

Before photosetting tests are started, the developing machine must be properly set up. In order to achieve optimum development the film and chemical manufacturers recommend a particular combination of temperature and development time. If development is carried out at 35°C, then the development time should, for example, be 30 seconds (refer to the manufacturer's guidelines!). Here the time between the

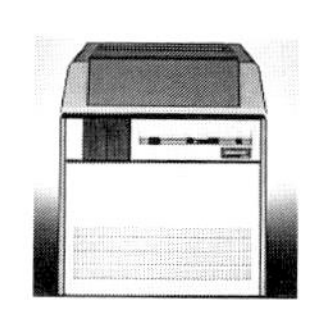

input and output of the film from the developing machine is clocked. Care must be taken to ensure that a good fixing takes place, i.e. that the temperature is not too high and that the time taken for the film to pass through is not too short, as this may lead to a high basic veil of over log 0.06. It is also recommended that the temperature display of the developing machine is checked.

Maintenance of consistent developing

In order to measure the developing process exactly and keep it constant, pre-photoset control strips are available from Kodak (Phototypesetting Control Strip). With these, the control field should reach a certain density. If the developing machine has been tested, a control strip should be developed and evaluated by means of a high-quality densitometer several times a day. If the system remains stable, it is sufficient for a control strip to be developed only once a day. The deviation in the daily production must not exceed log 0.05 to 0.1.

Photosetting testing

In order to test the photosetter a test file must be photoset with various light intensities. The light setting for special films should be selected so that no changes in half-tones occur during photosetting. The full-tone density should not be below log 2.6. Small deviations may be corrected using the fine setting of the photosetter intensity.

How does a photosetter work?

The so-called photosetter for DTP applications comprises

- an RIP (Raster Image Processor = PostScript Interpreter)
- the actual photosetter and
- the developing machine.

The photosetter photosets light-sensitive paper or film. It may be controlled by the computer using special commands. However, as PostScript should be utilised and not photosetter commands, an additional device, i.e. the RIP, is

RIP from Linotype

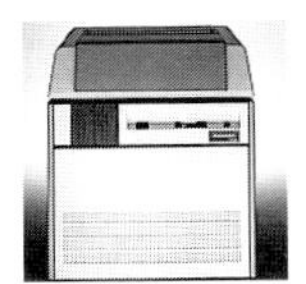

needed. When mention is made of photosetters in this book, the complete system including RIP, photosetter and developing machine is of course implied.

RIP (Raster Image Processor)

The RIP is a computer which processes PostScript commands for the photosetter. From the formulae and instructions which are contained in PostScript, pictures, text and lines are formed from the individual pixels. This pixel picture is then sent to the photosetter. Since about 21 000 pixels in the x axis direction and about 29 000 pixels in the y axis direction may be set for an A4 page, the RIP needs to be capable of considerable computation. Before the calculated pixel image is sent to the photosetter, the page is compiled in the RIP.

The largest allocation of computer time and effort is taken up by greyscale pictures. How quickly pictures can be photoset depends mainly on the RIP used. Older RIPs require much longer than newer versions. And even here, further developments will certainly cause photosetting costs to become more reasonable.

The RIP versions vary mainly in the speed with which they prepare PostScript commands and send them to the photosetter. A high-performance class of RIPs is offered with version 4 and higher from the Linotype company, through which greyscale pictures may be photoset quickly and hence also at reasonable cost. However, not only is the performance of the RIP a crucial factor, but also the speed at which data is transferred from the personal computer to the RIP. Connecting a PC to the RIP via a serial interface is less suitable for the processing of large volumes of data. Better results are obtained using a parallel (Centronics) or Ethernet interface. It would be even better if disks containing the Postscript data could be inserted directly into the RIP.

Photosetter: Linotronic 300 from Linotype.

If you agree either on fixed costs or costs depending on the size of the file for photosetting with your photosetting studio, then it will not matter to you which configuration of devices the studio uses. However, if the studio calculates costs on the basis of the time taken for photosetting, you are naturally at a disadvantage if it uses a slow device configuration.

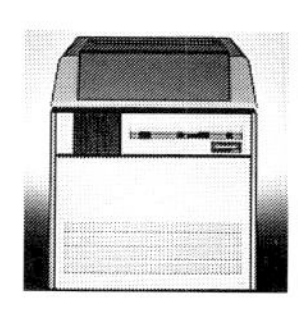

Photosetters

The pixel image which is sent from the RIP to the photosetter is next transferred to photographic paper or light-sensitive film by means of a laser beam. Just like the laser printer, the photosetter can only produce pixels of the same size. As the size of the dots cannot be changed, the individual raster dots for greyscale pictures must also be formed here from several pixels. The diameter of a pixel in a photosetter is about 0.02 mm!

Developing machine

Once the photosetter has completed its task, the photographic paper or light-sensitive film can then be developed. The paper or film roll is placed in the developing machine and after a few minutes the finished result is produced.

Overview

The following tables show the characteristics of several laser photosetters with their recommended raster widths.

Laser photosetters from Mannesmann Scangraphic

System[2]	Resolution		Set raster		Max. width
	dpi	pixels/cm	lpi	L/cm	(photosetting width)
Scantext 2030	1626 dpi	640 P/cm	[1]		300 mm
(helium–neon laser)	3251 dpi	1280 P/cm			
Scantext 2051	1626 dpi	640 P/cm	[1]		510 mm
(helium–neon laser)	3251 dpi	1280 P/cm			

[1] The manufacturer's settings were not entered in this table, as these photosetters are not preset uniformly.
[2] The laser photosetters from Mannesmann Scangraphic are also available as models with infrared laser diodes.

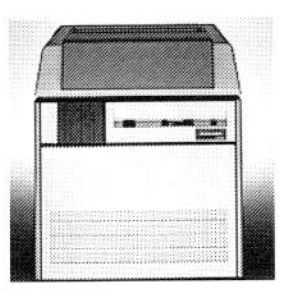

Laser photosetters from Linotype

System	Resolution		Preset raster[1]		Max. width (photosetting width)
	dpi	pixels/cm	lpi	L/cm	
Linotronic 300[2]) with RIP 3 and 4 (helium–neon laser)	635 dpi	250 P/cm	75 lpi	30 L/cm	305 mm
	1270 dpi	500 P/cm	150 lpi	59 L/cm	
	2540 dpi	1000 P/cm	150 lpi	59 L/cm	
Linotronic 330 with RIP 30 (helium–neon laser)	635 dpi	250 P/cm	75 lpi	30 L/cm	305 mm
	846 dpi	333 P/cm	75 lpi	30 L/cm	
	1270 dpi	500 P/cm	100 lpi	39 L/cm	
	1693 dpi	666 P/cm	120 lpi	47 L/cm	
	2032 dpi	800 P/cm	140 lpi	55 L/cm	
	2540 dpi	1000 P/cm	150 lpi	59 L/cm	
	3386 dpi	1333 P/cm	150 lpi	59 L/cm	
Linotronic 530 with RIP 30 (helium–neon laser)	635 dpi	250 P/cm	75 lpi	30 L/cm	457 mm
	846 dpi	333 P/cm	75 lpi	30 L/cm	
	1270 dpi	500 P/cm	100 lpi	39 L/cm	
	1693 dpi	666 P/cm	120 lpi	47 L/cm	
	2032 dpi	800 P/cm	140 lpi	55 L/cm	
	2540 dpi	1000 P/cm	150 lpi	59 L/cm	

1) The raster widths in this table are given by the manufacturer's settings, which, however, may be changed by the photosetting studio. The presetting is generally not required, as the raster width is set by the DTP software.

2) The laser photosetters Linotronic 300 from the Linotype company are the most common photosetters used. Linotype also offers the laser photosetter Linotronic 330 with a maximum resolution of 3386 dpi (1333 P/cm). Together with the new RIP 30 this laser photosetter is suitable for the photosetting of colour separations.

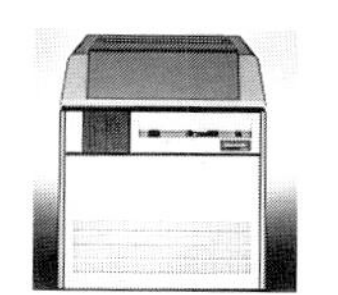

Laser photosetters from AGFA-Compugraphic

System	Resolution		Set raster[1]		Max. width
	dpi	pixels/cm	lpi	L/cm	(photosetting width)
SelectSet 5000 (helium–neon laser)	1200 dpi	472 P/cm	100 lpi	39 L/cm	408 mm
	2400 dpi	945 P/cm			
StudioSet 2000 (infrared laser diodes)	600 dpi	236 P/cm	100 lpi	39 L/cm	305 mm
	1200 dpi	472 P/cm			
	1600 dpi	630 P/cm			
	2400 dpi	666 P/cm			
ProSet 9400 (infrared laser diodes)	1200 dpi	472 P/cm	100 lpi	39 L/cm	328 mm
	2400 dpi	666 P/cm			

[1] The raster widths in this table are given by the manufacturer's settings, which, however, may be changed by the photosetting studio. The presetting is generally not required, as the raster width is set by the DTP software.

In general the desired raster width should already be entered in the DTP software. The entering of the nominal value (rounded value) is sufficient. The RIP then uses the exact value itself. However, entering the raster width may also be omitted, indeed sometimes it is not even possible to enter it at all, for example, in the case of grey underlining of text. In these cases photosetting takes place with the preset raster width of the RIP. In the tables above the preset raster widths of the manufacturers are given, which are the most common presettings found. Be wary since the photosetting studio may have set other values. It is best to consult your photosetting studio to ascertain which raster widths are pre-set, or check the settings with the aid of a raster counter.

On the basis of an example using a laser photosetter with 2540 dpi resolution, the following table shows that not every raster width may be set, or rather that certain raster widths are recommended.

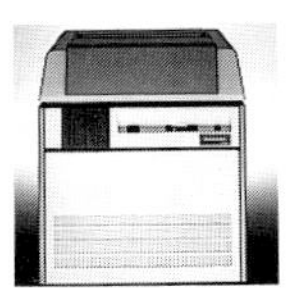

Recommended raster widths at 2540 dpi resolution		
	Raster angle 45°	
Actual raster width (of RIP)	Setting in the DTP software	
	Nominal value	Possible setting range
149,7 lpi	150 lpi	144 bis 156 lpi
138,2 lpi	140 lpi	143 bis 125 lpi
128,3 lpi	130 lpi	124 bis 133 lpi
119,7 lpi	120 lpi	116 bis 123 lpi
112,3 lpi	110 lpi	109 bis 115 lpi
99,8 lpi	100 lpi	98 bis 102 lpi
89,8 lpi	90 lpi	88 bis 92 lpi
85,5 lpi	85 lpi	84 bis 87 lpi
81,6 lpi	80 lpi	80 bis 83 lpi

Further specifications and tables on the recommended raster widths may be found in Chapter 12.

The following raster widths are usually used:

- 80 lpi (31 L/cm) for daily newspapers
- 120 lpi (47 L/cm) for reproduction on to paper of not very high quality
- 150 lpi (60 L/cm) for high-quality reproduction on to photographic printing paper

What is PostScript?

As PostScript is mentioned so often in DTP applications, we will briefly explain here what is so important about it. PostScript is a page description language which is device independent. It is a programming language with comprehensive graphics capabilities. With PostScript the structure of a printed page may be described (programmed).

Why then is Postscript so important to the DTP user? (We do not want to program and in any case definitely not using PostScript.) First of all PostScript has the major advantage that the structure of a page may be completely described. Any figures made up of lines and curves may be created, outlines may be filled with colour, text may be handled as graphics and greyscale pictures may be rasterised. All the necessary commands are available to create text, line drawings and photographs on a page.

If the DTP program, the laser printer and the photosetter (RIP) all speak the same language, the data can then be freely exchanged. A PostScript file may be printed by a laser printer or photoset by a photosetter. Exactly the same page layout is produced – only the quality varies. It does not make any difference whether the printer is supplied by company X or Y, as long as it understands the same language, i.e. PostScript.

The DTP program, aided by the PostScript programming language, thus creates a second program to describe the structured pages (text and pictures), i.e. the PostScript file. As the printer also understands PostScript, the printer or photosetting system, aided by the integrated RIP computer, then converts this program back into correctly arranged text and pictures. PostScript is therefore not dependent on the hardware. Printers from any manufacturer may be used, as long as the printer has the same computer program (PostScript interpreter) at its disposal. Only in this way can files printed on the laser printer also be sent for photosetting.

An important advantage therefore is the fact that PostScript is independent of the hardware and accepted as the standard by many manufacturers. A further advantage is that, as well as figures, text may be handled as graphics. This means that in the PostScript printer the pixel matrix for the individual letters is not described by a table, but instead a formula exists simply for the current letters or characters. Therefore, a PostScript printer only needs to store one size of character for each letter and then all characters may be enlarged as desired without loss of quality. Each character is created from a stored formula.

Hence, as text is handled as graphics, it may also be processed using other graphics programs (e.g. Corel Draw). A paragraph, a line or even individual letters may be changed, distorted, rotated or, for example, represented by various grey levels. The possibilities which are available using PostScript are diverse. For example, in Ventura Publisher a structured page may be saved as a PostScript file in EPS format (Encapsulated PostScript). This EPS file may then be reloaded into Ventura Publisher and enlarged or reduced by any amount.

PostScript printers, however, have the disadvantage that they are relatively expensive, since the complex printing programs need to be interpreted within the shortest possible time. These printers must therefore include very powerful processors. As

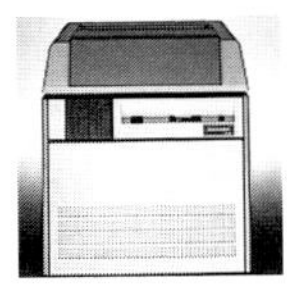

PostScript is relatively slow, when buying a printer you must also consider the speed of the PostScript interpreter, which is often not specified by the printer manufacturers.

A little history of PostScript

The language PostScript was developed by John Warnock and is the result of nearly 10 years of experience with various predecessor languages.

In 1976 in a research project John Warnock examined the application of interpreting languages and soon developed a language for CAD applications. In 1978 he went to the Xerox Palo Alto Research Center where he researched the 'black art' of photosetting, printing and graphics and where he developed the page description language Interpress. In 1982, together with Chuck Geschke, he founded Adobe Systems Incorporated and then released the third version of the language, i.e. PostScript.

Chapter 9
Printing

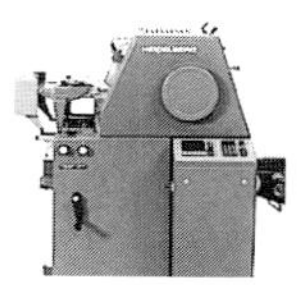

A new encounter

Before we rush into the subject of printing, we again need to make sure that we understand what DTP actually means.

When printing by the conventional method, texts and pictures were sent to specialists and experts in the printing industry made the best of them. The producer of the camera-ready copy, or the reproduction photographer, rasterised the original (paper picture, negative, slide, etc.) and then produced a film. The text was typeset and cutting and pasting was carried out, etc. Finally, when the complete film was ready, the printing firm carried out its task. Thus, in the conventional printing process, text and pictures represented the interface to the printing industry. You as the customer passed on your text and pictures and afterwards then criticised what you did not like. You needed no background knowledge as to how the final printing result came to be produced.

It is completely different with DTP. You are responsible for everything yourself, right up to producing the finished film. Instead of only photographs and text, you deliver the finished film. The printing firm is only required for reproduction. A bad original film can hardly be improved here. The original film (camera-ready copy) thus represents the interface to the printing industry.

You should note what this means. You now have the complete responsibility. If you desire the result of your work to be acceptable, you need to understand some important things about the printing industry. What the professionals previously undertook for you, you must now carry out yourself.

However, the experience of the experts in the printing industry cannot easily be transferred to the world of DTP. This means that these experts cannot readily support you when you encounter problems. However, even the manufacturers of DTP components cannot generally help you, since they do not understand enough about printing.

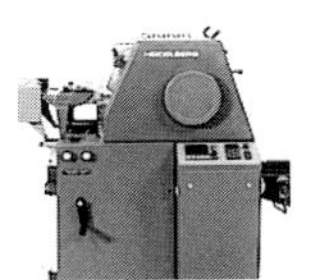

We have actually experienced it. The difficulties in comprehension between the DTP user and the printing experts are enormous. This is understandable as the printers simply receive the finished film and need only make sure that printing is carried out satisfactorily. Even the lithographer will not be of much help to you if he or she does not know the DTP system well. The personnel from the printing industry may have considerable experience in creating perfect work, but it is not easy to convert this experience into practical numbers for your computer. This short introduction is intended to illustrate how important it is to become familiar with several topics from the printing industry.

Quality with DTP

First we must expressly point out that, with respect to quality, the printing results created using DTP are in no way inferior to those created conventionally. The subject of printing covers many issues. Printing may be carried out on an impact, ink jet, laser or thermal printer. Even a photocopier represents a type of printer. We may have our printing carried out by a copying shop, in small offset or in offset printing. Relief printing (book printing, flexographic printing), gravure printing, planographic printing (offset printing), collotype printing and screen printing also exist.

The most diverse materials may be printed on. A description of various paper qualities would alone fill a complete book. Printing may be carried out in four-colour printing, in black and white or with spot colours (special colours). When pictures are printed criteria such as paper quality or tone increase need to be considered. Subjects such as quality control and standardisation are above all important in DTP. Printing can therefore turn into an open-ended subject. However, we do not want to train you to become a printing technician here, but instead we will limit ourselves to the most important issues which are important to you as a DTP user.

In printing instructions the format, number of sides, print run, paper quality, cost and deadline may be specified numerically. But what about the printing quality? Unlike many industries which work to specific standards, guidelines and tolerances, many printers attempt to print only 'as well as possible'. Amongst other effects, because of the use of various printing colours, printing cloths, machine types or different machine settings, the results produced from the same camera-ready copy may be completely different. Different values would need to be applied for the creation of camera-ready copy for each different printer. If a lithographic company creates the camera-ready copy, then these professionals have the necessary experience with their preferred printing firm and so are able to assess the actual printing result with their specially trained eyes. For a professional printing result it is essential that a test run be carried out beforehand, i.e. only a few samples are printed from the camera-ready copy so that the potential quality may be judged. If the result is not acceptable, either new camera-ready copy is created or the printers try again, within their capabilities, to achieve an optimum result.

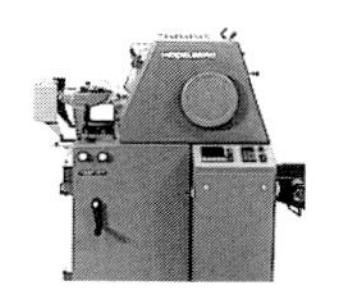

This process cannot be expected of a DTP user. Such a user does not usually have the necessary experience to be able to assess the camera-ready copy and subsequent printing process. In addition, experience is lacking regarding what to do with poor camera-ready copy to obtain a better result. Furthermore, working with DTP should considerably reduce costs and printouts since repeated photosetting is expensive. Just as important as the costs is the time factor. Furthermore, you probably do not want to 'limit' yourself to using only one printing firm, but would prefer to create camera-ready copy which various printers are able to print with the same quality.

Since the most common complaint is the lack of correlation between the quality of the camera-ready copy and the printed result, only clear rules and consultations will rectify this. Here we would like to explain these rules as well as the problems which are associated with printing. We will show that, through the tips and background information contained in this book, a DTP user may attain superb-quality pictures without lithographic experience. Lithographers will probably shake their heads in dismay at this point. We know the arguments put forward by the printing experts. They evoke the ideal printing result in discussions and make you aware of the most diverse problems, not readily grasped by the layman. These experts often overlook the fact that they themselves do not maintain the quality standard, mentioned in the discussions, in everyday operation.

We are not concerned here with creating a perfectly illustrated book through the aid of several numerical tabulations, but instead we would like to create a basis from which the DTP user may print standard publications incorporating high-quality pictures. You will find out that by using DTP good or even better results may often be obtained than with the conventional approach. At this point we would like to point out that in this book we are considering the printing of black and white pictures only. Of course, it is also possible to print colour photographs using DTP. At the present time the technology is being perfected (higher photosetting resolution and accuracy, various forms of dots, angles of rasters, new PostScript versions, colour calibrations, etc.) in order to attain perfect-quality colour pictures using DTP. However, this is a separate subject.

What is light?

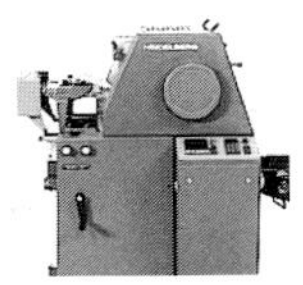

Even though we deal mainly with black and white pictures, we would like to mention briefly the subject of light, colour and human sight to provide a better understanding of the overall nature of the problems.

Let us for once start right at the beginning and consider what light actually is. Light consists of electromagnetic oscillations in a particular wavelength range. They are the same electromagnetic waves as gamma rays, X-rays, heat rays or radio waves. The physical difference between them all is in their wavelengths and consequently the various receiver devices. Even though they are essentially the same rays, in one case they are dangerous, in another we can see them and in yet another case they warm us. Light therefore represents only a small part of the total spectrum of electromagnetic oscillations (waves, frequencies). We are able to perceive different parts of these oscillations as different colours and the sum of all the visible frequencies appears to us as white light.

The following diagram shows the spectrum of electromagnetic waves. The visible range for the human eye starts at the end of the invisible ultraviolet rays at a wavelength of 400 nm (0.0004 mm, 8 × 1014 Hz) and ends at about 750 nm (0.000 75 mm, 4 × 1014 Hz). The enlarged scale section of the diagram shows the most important spectral colours.

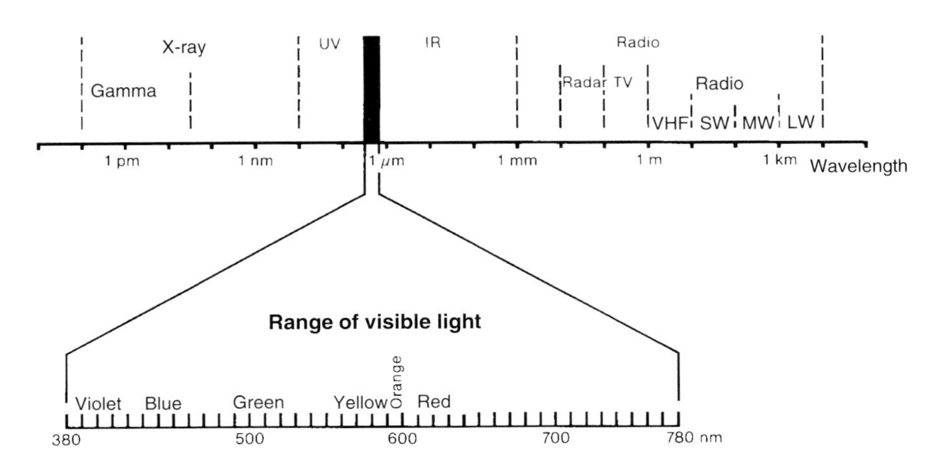

The spectrum of electromagnetic waves. Only part of the range is visible to the human eye.

How do we see light and colours?

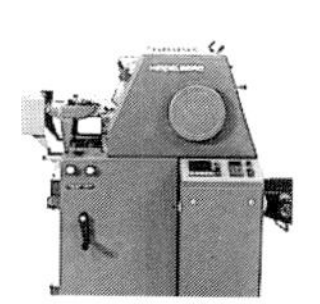

In the human eye there are three types of receptors responsible for our perception of colour and these vary in sensitivity depending on the wavelength. One of these receptors is receptive to violet–blue, one to green and the other to orange–red. If the receptors are not irradiated, black is perceived. If all the receptors are irradiated by the same amount, grey to white is seen. The brain therefore mixes the irradiances received from the eye (violet–blue, green, orange–red) additively to white.

An object absorbs part of the light shone on it. The eye perceives only the light reflected from the object. Hence the selection of a light source for the scanning or evaluation of pictures is important. If a particular colour, i.e. a particular frequency range, is missing from the light source then this colour cannot be seen. This is important in connection with the scanning of colour pictures, even if they are to be printed in black and white. Suppose you have a picture which consists mainly of red colours. If your scanner cannot recognise the colour red, then the picture is read into the computer incompletely and cannot subsequently be printed as a black and white picture true to the original.

The brightness of a colour determines both the raster tone and the colour density. The density specifies how much light the colour absorbs. The denser a colour layer, the less light reflected, i.e. the darker the colour tone becomes. When pictures are printed the various colour tones or grey levels are not achieved by various colour densities, but instead by means of the raster. When rasterising the composition of the colours is always the same. Various colour tones or grey levels are achieved by generating dots of various sizes, so that it is not the density of colour but the residual white area which determines the colour tone or grey level. The density (brightness) either of an unrasterised area (full-tone colour) or of a rasterised area may be measured.

Tone, brightness and density in principle are thus the same thing. The quantity that varies is the measurement unit. Tone is measured as a percentage and density on a logarithmic scale. The tone of a rasterised area (raster tone) determines the percentage cover of raster dots to white background, i.e. the grey level of an area. A full-tone area (unrasterised colour area) is given, not as a percentage tone, but instead as a density on a logarithmic scale. The density for a full-tone colour is dependent on the thickness of a colour layer and determines the intensity of the reflected light.

Methods for the creation of colours

In the following section, the two different systems for creating colours, i.e. the additive and the subtractive mixing of colours, will be explained briefly. The additive mixing of colours is wellknown to us from the colour monitor or colour television. By means of the three light sources red, green and blue (RGB model) virtually any colour can be created. If all three colours shine with the same intensity, white light is produced. This is readily observed on the television screen when the colour dots are examined closely. Colour pictures are described in television by means of the colour type (chrominance signal) and the colour intensity (luminance signal). The characteristic of additive colour mixing is that, by mixing three light sources appropriately, the desired colour is produced.

Subtractive colour mixing also exists and is wellknown to us from colour photographs. A colour picture is illuminated with white light (daylight, lamp). Owing to the colour layer of the picture particular light components are absorbed, so that only the colours remaining are reflected. Hence particular colours are subtracted from the white light.

In the printing industry colours are created by mixing the colours yellow, cyan (blue–green) and magenta (purple–red). This process is known as the YCM model. For reasons of quality the colour black is added, as a deep black cannot be created with the three basic colours. However, a process also exists in which black is used as the main colour. Hence the characteristic of subtractive colour mixing is that particular colours are created from white light by removing certain colours. Specifications for colour monitors or colour printers which claim that they can create 16.7 million colours are easily explained. By mixing three colours almost any colour tone can be produced. A resolution of 8 bits for each colour provides 256 levels for each of red, green and blue, which results in 16.7 million colours (i.e. $256 \times 256 \times 256$).

Colour separation

When considering the printing of colours, the term colour separation is often used. This may have several meanings:

- Spot colour

  Colours known as spot colours are not created by mixing colours, but are instead created by adding the required colour tone at printing time. When spot colours are printed separate camera-ready copy is required for each colour. If seven different colours are to be printed, then seven different films are required for each page. Colour separation here means, that the individual colours used are filtered out and photoset on to separate films.

- Four-colour printing

  Colour separation in four-colour printing means that the colours used are split into their primary parts, so that four films are required. 16 million colour shades are possible by mixing three primary colours. The fourth colour, black, is used to improve the quality. By means of a deeper black depths are created which could not be attained alone by mixing the three basic colours. The individual dots need to be arranged precisely so that moiré patterns are not produced. Raster widths and angles must be set differently and precisely for each colour. Very high demands are made on all components. For example, the photosetting studio requires special photosetters with high accuracy located in air-conditioned rooms. Finally, for the photosetting of films for four-colour printing new rasterising techniques (e.g. alternative dot formats) are being developed.

Colour copying with the computer

Although only black and white printing is covered in this book, we would nevertheless like to digress briefly to the printing of colour pictures. However, here we do not intend to describe the standard production of four-colour separated films by means of photosetters, but would like instead to describe a completely new method: colour copies by means of the computer.

As its latest development Canon, for example, is offering a colour photocopier (CLC 500), which can be controlled from a computer. It is therefore possible to produce quantities of about 240 A4 pages per hour immediately after laying out the document. Photographs and text are output with relatively high quality on normal paper of sizes up to A3. Anyone who requires only small print runs will much appreciate this advantage of being able to print immediately using the computer without inconvenience. Presentations, brochures, company newsletters, etc., may be produced with a speed and quality not achieved hitherto.

The colour photocopier is not only used as a reproduction device from the computer, but also behaves as a normal colour photocopier and may even be used as a scanner. A colour photograph may be placed in the photocopier, scanned and then transferred to the computer. Once the layout has been created using a DTP program,

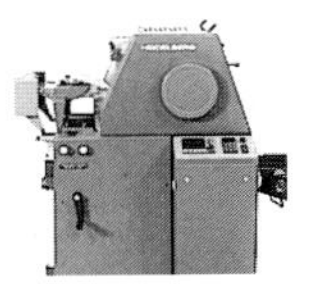

the document file is printed by means of the colour photocopier at a resolution of 400 dpi and the task is complete. Of course, pictures which have been obtained using normal DTP scanners or a video camera may also be printed.

The reproduction side of the colour photocopier functions, in principle, initially as a laser printer. As with a laser printer this colour photocopier houses a picture drum which is likewise sensitised by means of a laser beam. The difference is not that several colours may be overprinted, but that the size of the laser beam may be changed, allowing pixels of various magnitudes to be produced. A standard laser printer cannot vary the size of its pixels and so requires 8 pixels for 64 grey or colour levels. The colour photocopier, however, requires only one.

Also, as it is capable of enlarging by 400%, even large posters (about 1.1 m × 1.6 m) may be produced. However, the A3 sheets need to be pasted together since the output format is limited to A3. Since this colour photocopier is quite expensive, it

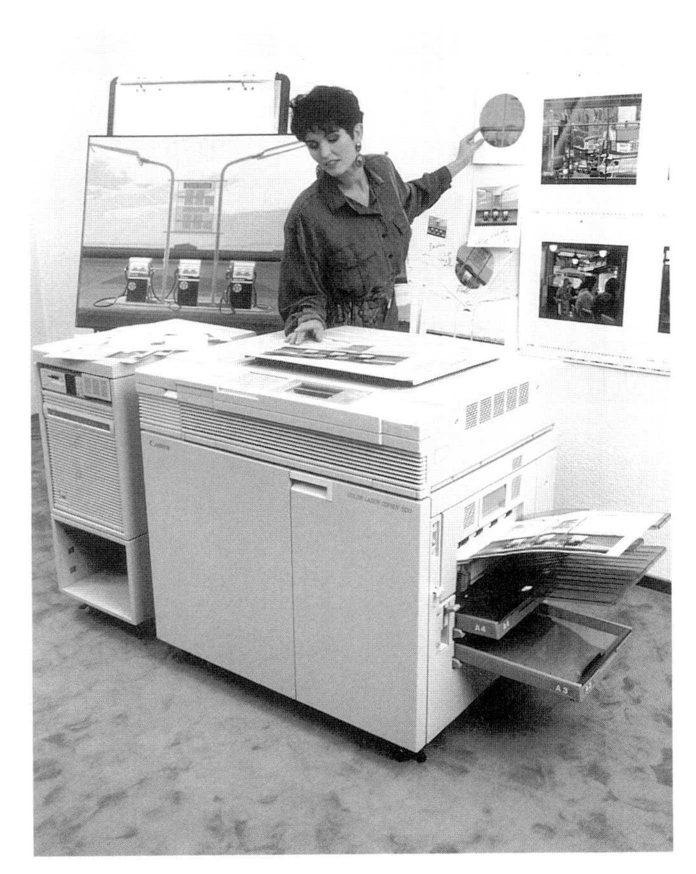

Colour photocopier from Canon, which understands PostScript and has an integrated scanner.

remains to be seen whether printers, copy shops or photosetting studios will use it to offer a suitable service to the DTP community. The potential is in any case fascinating: scanning, layout and printing – all in one day.

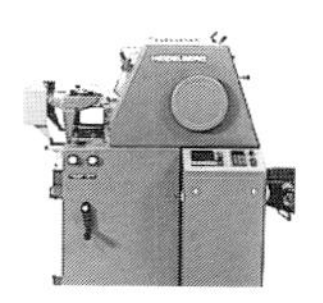

The special laser printer

Until now any mention of a laser printer implied a standard printer with a resolution of 300 dpi. Of course laser printers also exist with higher resolutions, whereby a correspondingly higher quality picture is obtained. In order to obtain high quality pictures the resolution should be greater than 800 dpi. PostScript laser printers with a resolution of 1200 dpi × 600 dpi produce pictures of good quality. However, these high resolution laser printers have the disadvantage of being very expensive and printing takes considerable time due to the large requirements of computer time and power.

However, a new variety of high resolution laser printers now exists, namely a PostScrip laser printer with DMV technology (Dot Multi Variation). Using DMV technology a picture may be obtained of a quality which corresponds to that of a resolution of about 2400 dpi. This is achieved because the size of each individual dot of a 300 dpi printer is changed by a factor of 16. From calculation this produces a resolution of 4800 dpi (300 dpi × 16 dpi = 4800 dpi). Since even professional photosetters only offer a resolution of 2540 dpi and it does not make sense to have an output quality of 4800 dpi on a laser printer, the DMV resolution for greyscale pictures is limited to 150 dpi (150 dpi × 16 = 2400 dpi). This results in a raster width of 150 lpi in the case of about 256 grey levels and may be produced even on a laser printer! In addition, these printers are able to attain very high output speeds, as time consuming calculations for the raster dots are no longer necessary.

The raster dots (and hence also the pictures) are created without losing any time during printing. As a result it is possible to output a photograph in less than 10 seconds. A DMV card is built into the PC and a driver is installed in WINDOWS for the control of the printer. As the processing power of the PC can be utilised and no additional processing power or memory extension is required in the printer, the DMV printing system is relatively inexpensive. If the printer is installed in a network, the individual workstations only require a WINDOWS driver in order to output pictures at a resolution of 2400 dpi and text at a resolution of 1200 dpi.

The news is however disappointing for Macintosh fans, since the DMV driver only functions with WINDOWS programs which output true grey levels (e.g. WinWord, Ventura Publisher, PageMaker, etc.)

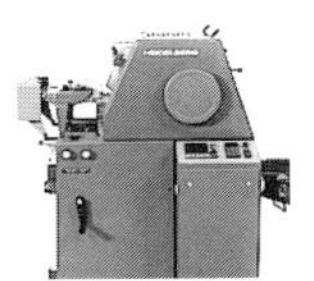

Offset printing machines connected to computers

Before we concern ourselves again with the actual subject of printing pictures and its associated basic problems, we must not forget offset printing machines connected to computers.

In the previous section, we introduced a colour photocopier with computer connections. At the present time, however, an offset printing machine is also available from Heidelberger Druckmaschinen AG which may be controlled directly from a computer. For the DTP user, this means that the intermediate step of 'creation of camera-ready copy' may be completely omitted. For the printing firm it means that printing plates no longer need to be produced from film. Therefore, basic costs are substantially reduced. This offset machine does, however, require considerable planning beforehand, as manual changes or film mounting are no longer possible.

Digital printing machines are available as four- or five-colour machines. Instead of a wet process, electrodes are used which write to printing film by means of electrical discharges. The printing film is made of polyester and covered with a thin aluminium layer. The electrical discharges from the electrodes puncture the aluminium, exposing the polyester. The polyester which has been exposed absorbs colours, i.e. printing occurs at these places. The aluminium layer rejects colours and so remains white. The printing machine at present attains a resolution of about 1000 dpi. As for a photosetter the printing machine is controlled by means of PostScript.

Heidelberg GTO-DI

Densitometer

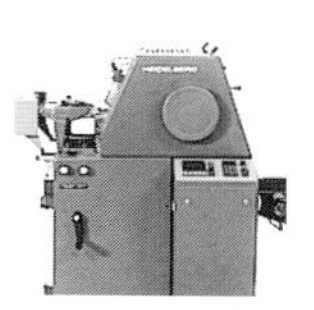

With a densitometer the brightness level of either a printed colour layer or a rasterised area may be measured (density or raster tone). The density determines how much less light is reflected from an area with respect to a completely white area. For this reason a densitometer is always calibrated before a measurement by means of the white area currently being used.

The densitometer shines light on an area and measures the strength of the light (brightness) reflected back. This light strength is converted into an electrical current and indicated on a display. This is the measurement of the ratio of white to the colour layer. The thicker the colour layer, the less light that is reflected, and hence the larger the difference to that of the white area and so the higher the density. However, in the printing industry the measured brightness is not given as a percentage, but is printed as a layer density on a logarithmic scale.

The printing expert, for example, uses the densitometer to measure the density (brightness) of colour (full-tone density). However, above all, the brightness of a rasterised area is measured using a densitometer. For example, it is possible to measure how much the 40% grey level on camera-ready copy changes during the printing process. Therefore, during the printing process, the tone may be controlled and, if necessary, altered. When printing, in order to measure colours which are still

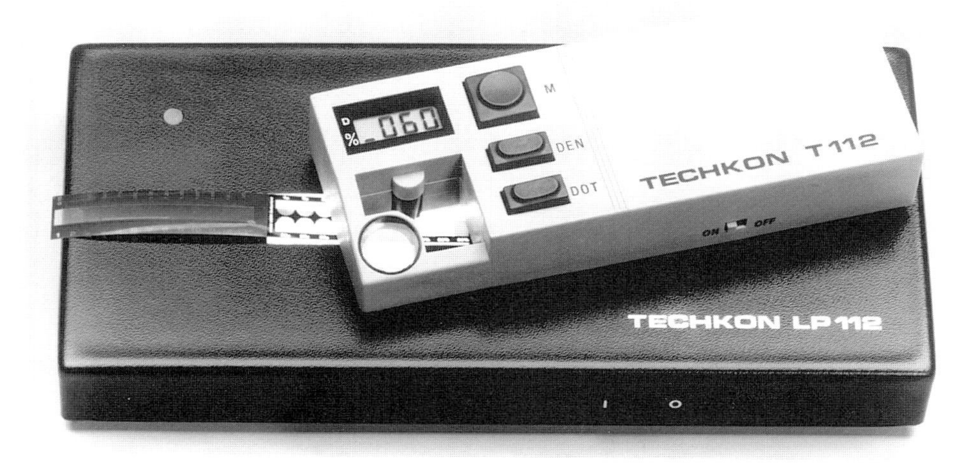

The density and raster tone are measured with a densitometer. Using a reflection densitometer a paper copy may be measured and with a transmission densitometer a film may be measured.

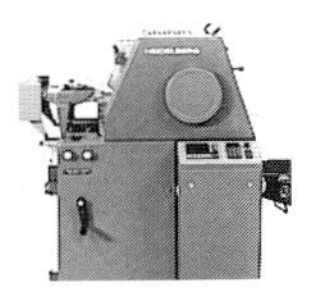

wet, or to measure high full-tone densities accurately, the densitometer must have a polarisation filter. So, if you obtain a densitometer, make sure that a polarisation filter is built into it. Otherwise, you will only be able to measure accurately the grey level of rasterised areas on film or paper.

The following example illustrates a logarithmic measurement scale with respect to density:

Logarithmic density	Greyscale (percentage)
0.01	2%
0.05	11%
0.1	21%
0.2	37%
0.3	50%
0.4	61%
0.5	68%
0.6	75%
0.8	84%
0.9	87%
1.0	90%
1.5	96%
2.0	99%

A table for converting density values into percentages may be found in Chapter 12.

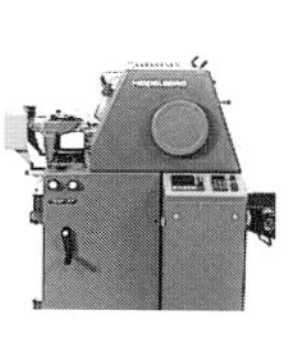

Which raster width does the eye prefer?

Before the various typefaces, paper qualities and resulting raster widths are introduced, we must first examine how we perceive the individual raster widths. We should develop a feeling for what different raster widths mean to our sight.

Two values influence the perception of raster images. At one extreme is the limit at which the eye is able to integrate individual raster dots to a grey level. At the other extreme is the limit where no more dots are recognisable. If dots are no longer recognisable there are no real advantages in increasing the raster width.

The limit at which grey levels are recognisable and the limit at which dots are no longer recognisable depend on the distance of observation. In the following we assume that printing results are normally viewed from a distance of 30 cm.

In a raster width range of 50 to 80 lpi (20 to 30 L/cm) pictures can be recognised, but the pattern of raster dots is very distracting. For raster widths of 80 to 100 lpi (30 to about 40 L/cm) the raster dots are no longer distracting.

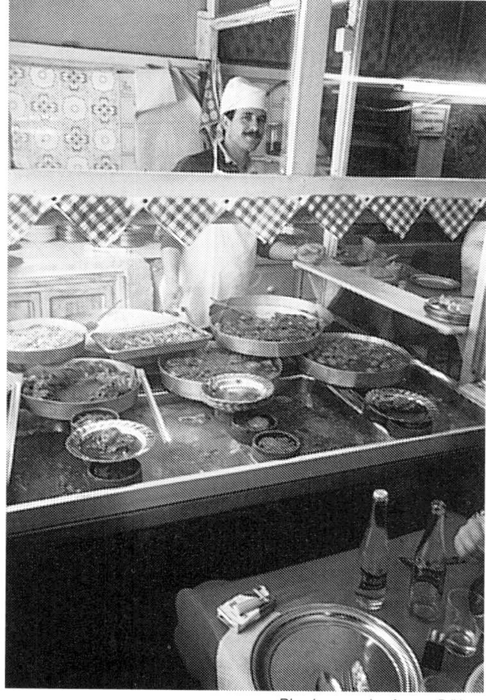

Photograph: Anne Böhm

Raster width 75 lpi (30L/cm)

Raster width 100 lpi (39 L/cm)

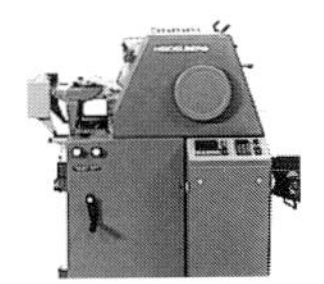

For raster widths from about 100 lpi (40 L/cm) the dots are hardly perceptible and at raster widths of about 130 lpi (50 L/cm) the eye is not capable of discerning the individual dots. At yet higher raster widths the resolution of detail may rise, but under normal observation conditions the eye is no longer capable of detecting this.

In addition, raster widths which are too high are not sensible because at high raster widths the contrast is reduced considerably and more problems may arise when printing. 150 lpi (60 L/cm) is the standard value for normal high-quality offset prints.

From the sight capabilities of our eyes particular raster widths are applied under normal observation of raster images (distance to the picture of about 30 cm):

- From 50 lpi (20 L/cm) — Picture recognisable, but raster dots very distracting.

Raster width 120 lpi (47 L/cm)

Raster width 130 lpi (51 L/cm)

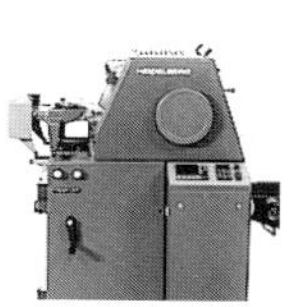

- From 80 lpi (30 L/cm) — Picture and raster dots easily distracting.
- From 100 lpi (40 L/cm) — Good-quality picture, raster dots no longer disturbing.
- 130 lpi (50 L/cm) — Good-quality picture, raster dots no longer visible, details very sharp.
- 150 lpi (60 L/cm) — Very good-quality picture, raster dots no longer visible even if viewed more closely, standard value for high-quality prints.

As the pictures demonstrate, it is not always necessary to use the relatively high raster width of 150 lpi (60 L/cm).

Raster width 140 lpi (55 L/cm)

Raster width 150 lpi (60 L/cm)

Various raster widths

75 lpi (30 L/cm)

100 lpi (39 L/cm)

From film to printing plate

Printing cannot be carried out directly using the normal camera-ready copy (film, paper) which you send to the printers, but instead a so-called printing medium (printing plate, printing film) needs to be produced. However, there are exceptions: so-called direct printing films may be used in a photosetter or laser printer and these may then be used for printing without further processing. The additional creation of printing media may subsequently be omitted.

It is not necessary for a DTP user to know about the different printing media. It suffices to know which type of camera-ready copy you can send to the printer. The printers choose the correct printing media depending on the actual camera-ready copy, the desired quality and the print run. However, in order to save costs and to attain an optimum degree of quality, it is nevertheless useful to be aware of the different printing possibilities. Essentially, printers may produce prints from any copy. However, if you deliver the correct copy to a suitably equipped printer there are immediate benefits.

Printing media made of aluminium

Usually, an aluminium printing plate is produced from the photoset film. A thin layer of light-sensitive material is placed on this aluminium plate. The film (camera-ready copy) is placed on top of the light-sensitive layer, with close contact between the plate and film achieved via a vacuum. The printing plate is then photoset so that an exact copy of the film is produced on the printing plate. The layers exposed to light are eroded and then cleaned away with developer. Finally, the printing plate is rubberised. The finished printing medium, namely the printing plate, is now available. This printing plate is clamped into the offset printing machine and printing can then begin. The layers which have not been eroded transport the colour during the printing process. The cleaned layers absorb water.

Whenever printing plates are mentioned here, it is worth while knowing that nowadays these are no longer heavy plates, but thin aluminium sheets with a thickness of up to 0.5 mm. Depending on the page size and type of printing machine several pages may be placed simultaneously on a printing plate. The creation of such printing plates produces the highest quality, but is also the most expensive process for small print runs. It is not so much the material cost which needs to be weighed up here, but rather the costs required for the make-up of the camera-ready copy and the developing, as well as for the assembly of the printing plate in the printing machine. The aluminium plate processed further is an expensive raw material.

As long as your printing result only has a small number of pages and a large number are to be printed, the basic costs should certainly not play an important role. However, if you create a manual, for example, which consists of many pages and needs to be updated continually, then the cost for creating the printing media must be considered carefully.

Economically from computer to plate

Processes do exist where the creation of printing media may be totally omitted. These are known as 'Computer-to-plate'. With a so-called direct printing film (Setprint film) printing may, as the name suggests, be carried out immediately without an intermediate step. Instead of the standard printing plate, Setprint film is used. A Setprint film is relatively easy to create and can be used instead of the regular film material in the photosetting studio. However, the photosetting studio does need to be equipped for the processing of Setprint film, since the developing process is rather different to the processing of normal film. On the other hand development is carried out continually. In the meantime, systems are already available where aluminium printing plates may be photoset directly from the computer.

Other printing media

Various types of printing media exist. You have already met the best-quality printing medium, namely the aluminium printing plate. In order to photoset the aluminium plate, a film must usually be available. The film is produced by the photosetting studio from data stored on a disk. However, it is also possible to produce a printing plate made of aluminium from paper camera-ready copy. Here, for example, the printers then produce the film or aluminium plate using a process camera by means of an electrostatic process.

As a further possibility, the so-called direct printing film, which has already been mentioned, is usually produced by the photosetting studio from data on a disk. However, direct printing film may also be produced from paper camera-ready copy. For example, a daylight plate camera exists through which a direct printing film may be created from paper camera-ready copy within about a minute. It is possible to print immediately using this direct printing film, without the necessity of an additional aluminium plate. Even direct printing film is often simply referred to as 'printing plates'. Direct printing films may be made from various materials such as paper or polyester. The simplest direct printing film is a paper film, which may be processed by a laser printer. When using special film printed on a laser printer, the printing firm can print immediately using a small offset printer without further media being required.

From tone increase to quality control

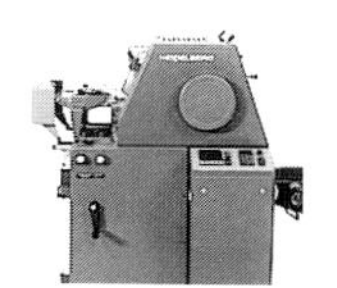

Next we will consider the most important topic regarding the printing of pictures. The characteristics of the printing process must be taken into account when scanning or processing the picture.

Pictures become darker due to the increase in tone

If a greyscale picture is to be printed it must be rasterised, i.e. the picture then consists of individual dots each of which has a particular size. Unfortunately, the individual raster dots become larger on printing, i.e. the tone increases and the picture becomes considerably darker. Hence reference is made to a 'tone change' or 'tone increase' when printing. Tone increase implies an enlargement of the individual raster dots.

In principle the tone may change for each 'copying process'. As already described, this may also be the case when photosetting. When film camera-ready copy is transferred to a printing plate another change in tone occurs. For example, when a positive film is copied on to a printing plate the tone may become about 4% lighter, or when a negative film is copied about 4% darker. When printing from the printing plate on to paper the tone becomes darker again. This is about 26% for a medium grey level. However, as the intermediate stages are not really of interest when producing a printing result (from film via printing plate to printing result), the specification of tone increase includes the creation of the printing medium.

Tone increase is defined as the difference in brightness between the film camera-ready copy and the final printing result.

It must be noted that, depending on the grey level (size of dots), the change in tone is variable. For example, for a grey level of 40% the tone increase may be 22% , but for a grey level of 20% it may only be 13%. Hence, unpleasant surprises may occur after printing. A picture may appear to look quite good on the monitor or on the film, but the printers then deliver a picture which is far too dark or flat. The amount by which the tone increases due to printing depends mainly on the type of paper used, the raster width and on the printing firm itself. As it is also dependent, for example, on the type of printing machine, setting of the machine, type of printing cloth, type of colour, thinner and wetting, etc., the increase in tone may likewise vary from printing firm to printing firm.

Considerable differences in tone increase are also due to the quality of paper used. The tone increase is greater in absorbent paper than in paper which is less absorbent (coated paper). The raster width also effects the tone increase. The tone increase is greater for many small dots (fine raster) than for dots which are larger (coarse raster).

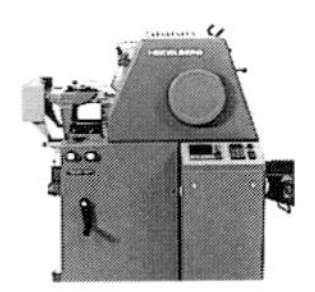

As pictures become darker during the printing process they must be made suitably lighter beforehand.

Hence, for example, when using paper of high quality a 40% grey level must be made 12% lighter if the tone is to be 40% again after printing. This means that you must take this tone increase into account when actually scanning or processing the picture. However, you can only do this properly if you know what quality of paper and raster width you are using. The quality of paper you decide upon therefore influences the type of image processing and maximum raster width which may be used.

In this case, image processing means that the grey levels need to be 'distorted' before printing. This 'distortion' may of course only be carried out, using an image processing program, once the picture has been set. If the picture is set to an optimum level on the screen, then the gradation (change in tone) needs to compensated for again before printing. Of course, a prerequisite for this is that the monitor was calibrated.

The higher the raster width, the higher also are the requirements for the printing process itself. Hence in general the raster width should be selected as low as possible, as then fewer problems occur when printing. The finer the raster, the greater effect errors in the creation of the printing medium, and subsequent printing, will have. Usually a raster width of 150 lpi (60 L/cm) is used for high-quality prints. However, raster widths of 140 lpi (55 L/cm) or 120 lpi (48 L/cm) are also used.

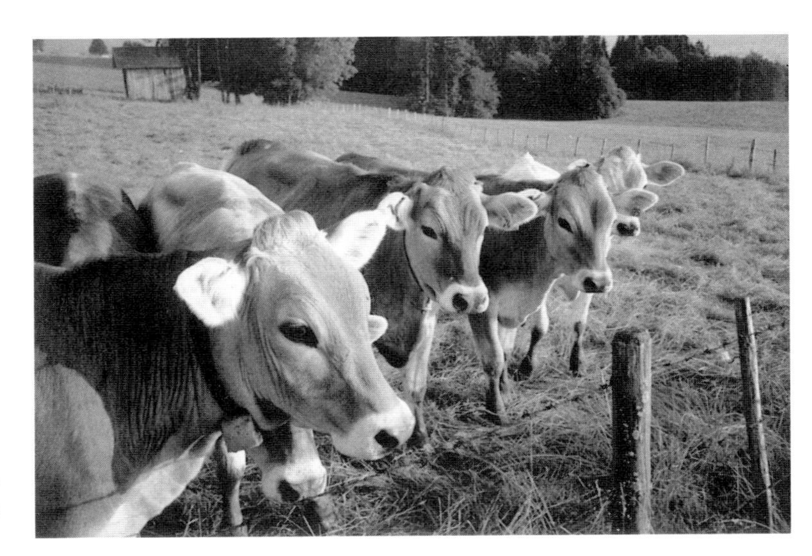

This is what the original picture looks like.

Light trap, raster tone, area covering

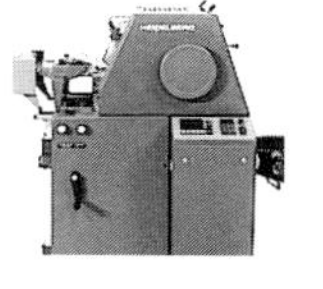

If you hold a piece of paper up close to a light, part of the light shines through it. This demonstrates that light is able to penetrate the paper layers. If paper is illuminated, part of the light is absorbed in the paper and part is reflected. The more light which is reflected, the lighter the paper appears to be. Light penetrates the paper layers, is scattered therein and finally emerges at the surface again. As raster dots lie on the surface of the paper, the individual raster dots prevent part of the light from escaping to the surface again. This effect is known as a light trap. Owing to the trapping of light, the picture also appears a little darker. The finer the raster, the darker the picture due to the light trap.

The combination of paper composition, raster width and printing process influences the quality of a printed picture. These effects must be taken into account beforehand when scanning or image processing, so that a scanned 50% grey level remains approximately 50% after printing. If reference is made to an increase in tone or in raster tone, or if the density is measured using a densitometer, then it is understood that the light trap has already been taken into account. The area covering only specifies which percentage part of the area is covered by the raster tone.

Standardisation in offset printing

Often it is not clear to us what standards actually mean. For example, if no standard existed for paper size, everyone would use a different one. Each printer would then require a special paper size, each requiring a different file to store it. The personal

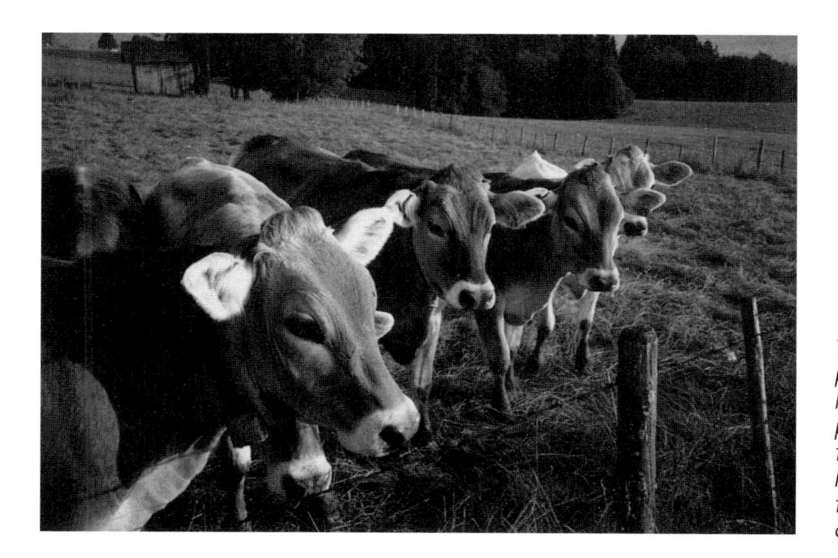

This is what the picture would look like after printing if the tone increase has not been taken into account.

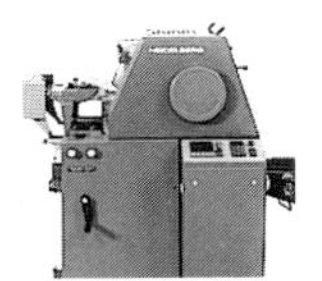

computer only became generally accepted, because its special design and function was adopted as a standard to which each PC manufacturer complies. In the meantime, the printing industry is also striving for standardisation, even though some printing firms are fighting against it. Previously it was possible to print precisely how one desired or was able to, whereas with standardisation exact values must be adhered to. This is to the advantage of the customer who consequently will know exactly what to expect. From each printing firm which complies with the standard values, you will obtain similar results.

Through standardisation it is also easier to control the work produced by a printing firm, as specific printing characteristics are then fixed. However, even the printing firms will have advantages. Fewer complaints will be received and the firms will be able to work more efficiently in so far as reproductions from various sources may be printed in one pass.

The DTP user, who does not have the required printing experience, relies on standardisation. It is only in this way that pictures can be scanned with uniform values and then merged into the DTP document. When working to standardised values, it is easy to scan, process and photoset pictures. Any printing firm may then be chosen, since it is bound to the standardised values.

FOGRA determined the values for standardisation. FOGRA is the German research association for printing and reproduction technology. However, research associations in Great Britain and the United States have independently obtained approximately the same results. The difference consists mainly in the specified measurement process and in the fact that FOGRA's standardisation has been published for all offset printing machines. The difference which occurred in the measurement processes was that FOGRA measured the tone increase for the 40% and 80% grey levels at a raster width of 150 lpi (60 L/cm) whilst other institutes measured the grey level of 50% at a raster width of 135 lpi (53 L/cm). If the results from the calculations for these measurement processes are scrutinised it is found that they are almost identical. Since the standardisation to FOGRA is widespread and has already been specified in detail, you should always refer to its values. Even for high-quality prints a raster width of 135 lpi, 150 lpi or 180 lpi is common, so that a raster width of 150 lpi is sensible for measurements. This is also confirmed by the fact that a maximum of 150 lpi should be used in many photosetters.

Of course, in principle, it is better if the tone increase in printing is as low as possible. However, if printing is carried out with a tone increase which is too low, the full-tone density will also be too low, i.e. black is then not deep black, but is instead too light. The pictures then appear to lack contrast. The art of printing therefore consists of attaining the highest full-tone density possible (deep black), without the tone increase being too high. Hence, in general, the tone increase recommended by FOGRA should be used for printing. However, what kind of quality is finally produced depends on the ability and equipment of the printing firm. In the end, they should print with the tone increase which they perceived when creating the film.

The tone increase is usually specified for a raster of 150 lpi (60 L/cm) and for the two grey levels of 40% and 80%. In general, for good-quality paper and to match the tone increases specified, a full-tone density of about 1.5 to 1.8 is attained for the colour black. A good printing firm may even attain full-tone densities of over 1.9. The full-tone density specifies the thickness of the colour layer for a 100% full area.

When creating the printing medium, and during printing itself, considerable changes occur in the tone. The standardisation to FOGRA establishes with which change in tone a printing plate may be created and how printing may be carried out.

The tone increase is the difference in tone between the camera-ready copy (film) and the printed result. Changes which occur when the printing plate is created are therefore included in these specifications. The FOGRA specifications of tone increase are with respect to the grey levels of 40% and 80% at a raster width of 150 lpi (60 L/cm) and, in our case, to the colour black. Other values are valid for colours.

The tone increase for a grey level of 50% and a raster width of 135 lpi (53 L/cm) is almost identical to the tone increase for a grey level of 50% and a raster width of 150 lpi (60 L/cm).

Standardised tone increases are shown in Chapter 12.

As a DTP user you should designate specific grey areas in your DTP document to be filled with grey levels of 40%, 80% and 100%, and also set a raster width of 150 lpi (60 L/cm). You can then measure these grey areas after photosetting and after printing, so that the 40% area, for example, may have a grey level of about 59% after printing. You must make sure that a raster width of 150 lpi is set for the grey areas. The 100% area is required since grey levels are always relative to 100% black. In addition, if the 100% area is measured, it is then possible to monitor whether the printing firm has achieved the desired deep black (full-tone density).

When carrying out measurements using a densitometer, it must first be calibrated with respect to the white paper. The grey level (e.g. 40%) and the full-tone density (100% black) are then measured for use as references. The densitometer then shows

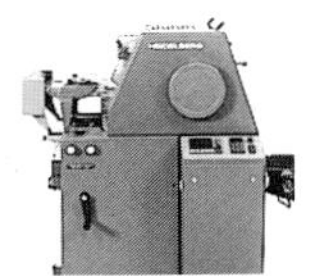

the grey level as a percentage. It is also possible to measure the density (logarithmic value for brightness) of a grey level and then to determine its percentage value by using tables.

It does not matter at which values your printing is carried out, but for a large amount of printing it is the tone increase and not the full-tone density which should be fixed. Controlling the full-tone density during further printing leads to worse results than if the tone increase in the 40% and 80% fields is controlled. If you print with the specified standard values you will have the advantage of printing with a uniform quality. It will also be easier to employ an alternative printing firm.

However, first find out what values the printing firm uses. If you are lucky, the printing firm will possess a densitometer and know the correct values. If it does not have a densitometer, you may still be able to achieve good results. However, in this case you should express your wish that a considerable amount of colour is added.

Offset printing practice

In the previous section we mentioned that the printing firm should comply with standardised values and that it should enquire about the tone increase. Unfortunately, however, we must add that this is usually not that easy. Many printers are not used to working with exact measurements. Often a picture is judged by eye and the printing process set accordingly. Of course, it may be that the printers themselves do not know the exact tone increase and hence the values may vary from printer to printer. It was also difficult for us to ascertain the tone increase from the printing firms. Sometimes, simply any values were given. These were only corrected, when they were to be written down in the contract. In addition, it is often only worthwhile for the printing firm to set the machine exactly for larger jobs.

In large print runs, and for prints of large sizes, a measuring strip can be printed at the same time for monitoring purposes and then disposed of when the prints are cut. However, a densitometric monitoring is hardly possible for the small offset print run, as the printing is carried out with the correct size paper and so cannot be cut to the required size afterwards. Printing should also be economical using small offset printing, so the higher cost and effort for quality controls are not acceptable. In addition, often it is only trainees who look after small offset machines and they do not possess the necessary ability to print high-quality pictures.

However, it is not only with printing that skill is required, since considerable errors may occur when the printing medium is produced. Especially in small offset printing the cost and effort should be kept as low as possible and so economical ways of creating printing media are also used. For example, if an X-ray unit is available, a printing firm will produce a printing film from paper or film camera-ready copy using a plate camera instead of an aluminium printing plate.

It is certainly also useful if you analyse the quote on the basis of the following criteria:

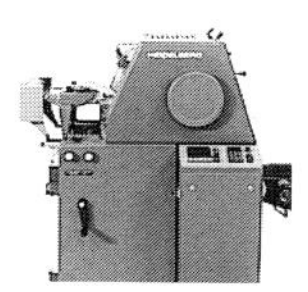

- Costs for the first print

  The costs for the first print include all the preparatory work which needs to be carried out before the actual printing can be started. On the basis of these costs you can estimate whether it is worth while choosing a different printing medium. For direct printing media no costs arise for the creation of a printing medium. However, modest costs will be incurred for the setting up of the printing machine. You must inform the printing firm which type of camera-ready copy (film, paper, direct printing film) you are going to send.

- Print run cost

  The print run cost includes all the costs for the desired number of copies. Here you should inquire about the prices for various qualities of paper. In addition, take the cover pages into account, since another quality paper may be needed.

- Cost for further processing

  Further processing includes, for example, the collation of individual sheets to form a newspaper. You must specify which form the end product will take. If, for example, the sheets are to be folded and stitched, then it is sensible to cut the document after folding so that the inner pages do not protrude by varying amounts.

- Other costs

  Other costs include, amongst other things, packaging and delivery.

Furthermore, both when requesting a quote and in placing the order with the printers, mention ought be made that printing should be carried out according to the values for the tone increase mentioned in this book. (Please refer to chapter 12 'Scanning and printing of pictures: summary and tables'.) The tolerances for the tone increase are usually interpreted as a 'standard deviation' in the statistical sense. This means that only 68% of samples from a print run need to lie within the specified tolerances. Smaller tolerances therefore need a separate agreement.

Which printing medium for which purpose?

This chapter is probably less interesting for most readers. You have your files photoset on to film and then send this film to the printing firm. In the end it does not matter to you which printing medium the printers use, as long as the quality is acceptable. However, if you want to save money, it can be useful to know about the alternatives available for the production of printing media. For example, it is possible

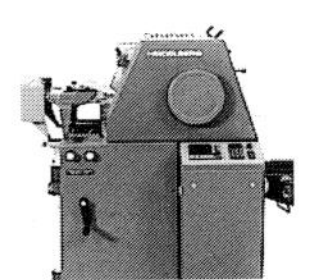

to photoset to direct printing film, or even to use direct printing film in the laser printer. With direct printing film the additional creation of printing media is not required. Of course it is possible to use any printing process with DTP. We are, however, only interested in typical DTP applications here. Many hybrid methods of creating camera-ready copy may be conceived which complement each other. Thus, for example, a colour photograph may be integrated by conventional means if no colour equipment is available.

The use of so-called spot colours is possible even without colour equipment. In this case individual parts (drawings, lines, test) are structured in colour. So-called colour separation is carried out here, where a separate film is created for each colour. This does not mean that colour photographs can be printed that easily using DTP. Photographs are produced mainly by four-colour printing, i.e. about 10 million colour nuances can be created by mixing the three basic colours. With this colour separation only four films (including black) are required in total.

The cost of the different printing processes varies mainly in the cost of creating the printing media. In a comprehensive document with a small print run these basic costs are of course more important than when two A4 pages are printed with a print run of about 10 000. For small print runs reproduction is also possible by means of a photocopier, though greyscale pictures, however, can only be reproduced with very poor quality. For this reason we will not consider this method further.

The selection of the printing medium is based primarily on the quality requirements and whether the printing result is to be bound, folded, stitched or glued. Good quality, with raster widths of more than 149 lpi (55 L/cm), may at present only be achieved through the use of aluminium printing plates.

The choice of printing medium also has an influence on how the individual pages are to be bound. The simplest method is where the individual pages have gummed backs. The pages are printed, sorted and glued in order. It is slightly more difficult if you wish to produce a document in which the individual pages are folded, laid inside one another, and, if necessary, stapled or stitched. For example, if you wish to create an A4 document, then the individual pages need to be printed as A3 pages and then folded. The individual A4 pages therefore need to be arranged in the correct order on the A3 sheets.

The arrangement of several pages on a printing medium is known as 'imposition'. The printers must ensure that they maintain the correct order of pages in this. If you send film or paper camera-ready copy to the printing firm and the printers then produce the printing medium (aluminium printing plate or printing film) from this, you do not need to worry about the imposition. The printers are responsible for arranging the pages correctly on the printing medium.

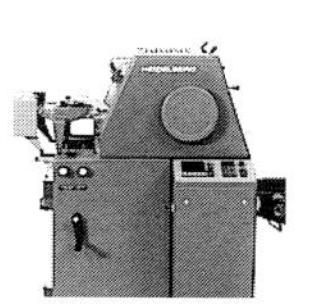

If direct printing film is used and the individual pages are not glued, but instead folded and laid inside each other, then you need to organise the imposition yourself. In every case you should first consult the printing firm or bookbinder.

When creating direct printing film the arrangement of pages must be correct right from the start. Hence the pages must be arranged and output from the computer in the correct order. In a four-page brochure, manual arrangement is still relatively simple. Pages 1 and 4 are placed on one sheet and pages 2 and 3 on the other. With a multiple page document this is no longer possible manually. In this case programs exist which send the pages from a PostScript printing file in the correct arrangement and order to the printer or photosetter. Hence a handy A5 leaflet may be created easily even on an A4 laser printer, without any cutting and pasting being necessary.

A newspaper may be produced as follows:

The number of pages is divided by four, yielding the actual number of sheets. Next, these sheets are laid out and folded together in the centre. The individual pages are then numbered. Finally the sheets can be separated again, making it possible to see which pages belong to which sheet. All odd numbers appear on the right of the binding and all even numbers on the left. The sum of both page numbers on the same sheet (right and left side) always produces the same value.

The following list shows the most important printing alternatives:

a) High-quality prints

Camera-ready copy:	photoset film
Printing medium:	aluminium plate
Maximum size:	72 cm × 102 cm and even larger depending on the printing machine
Print run:	small to very large print runs

b) High-quality text, good-quality pictures

Camera-ready copy:	photoset direct printing film
Printing medium:	not applicable
Maximum size:	A4 or A3
Print run:	up to about 10 000
Remarks:	The number given for the print run is valid for wood-free paper. For larger print runs it is usually more sensible to work with films and aluminium plates.

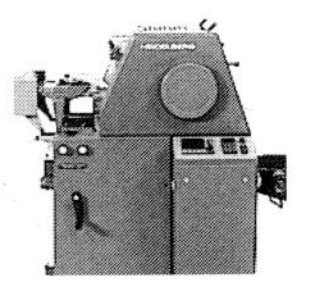

c) Good-quality text, good-quality pictures

Camera-ready copy for text:	paper offset film from the laser printer
Camera-ready copy for pictures:	film, direct printing film, photoset paper camera-ready copy
Printing medium:	not applicable
Maximum size:	A4 or A3
Print run:	up to about 2000
Remarks:	Pages which contain pictures need to be photoset. Pages containing text only are output by the laser printer. This process is only sensible if a large number of text pages and only the occasional picture need to be printed, or if the pictures are added in an appendix. The number given for the print run is valid for wood-free paper. For larger print runs several printing films need to be produced.

d) Good-quality text, very poor-quality pictures

Camera-ready copy:	paper offset film from laser printer
Printing medium:	not applicable
Maximum size:	A4
Print run:	up to max. 1000
Remarks:	This process is only appropriate if no greyscale pictures are to be printed. The number given for the print run is valid for wood-free paper. For larger print runs several camera-ready copies need to be produced.

The following table provides an overview of the various printing media which are available for offset printing.

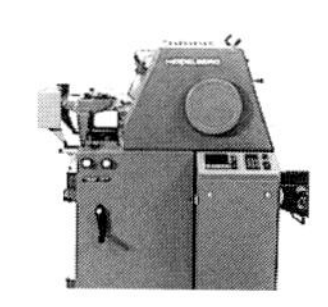

Printing media for offset printing				
Process	Camera-ready copy	Plate material [1]	Raster	Print run
Photosetting	Film, creation of printing medium necessary	Aluminium plate [1]	>150 lpi (59 L/cm)	> 30.000
Diffusion transfer (DTR)	Paper camera-ready copy, less suitable for DTP [2]	Paper, plastic, aluminium	120 lpi (48 L/cm)	Up tp 30.000
Electrostatic	Paper camera-ready copy, less suitable for DTP [2]	Paper with zinc oxide layer	120 lpi (48 L/cm)	Up to 5.000
Photosetting	Direct printing film, may be used immediately for printing	Not applicable	120 lpi (48 L/cm)	Up to 10.000
Laser printer output 300 dpi	Paper printing film, may be used immediately for printing	Not applicable	60 lpi (23 L/cm)	Up to 2.000
Laser printer output > 1.000 dpi	Paper printing film, may be used immediately for printing	Not applicable	100 lpi (40 L/cm)	Up to 2.000

1) *Various printing media exist: granulated aluminium plate for a print run of about 30 000, anodised aluminium plate for a print run of over 30 000 and for very small raster widths. In addition, so-called compound metal plates exist which may be used for extremely high print runs (over 500 000).*

2) *'Less suitable for DTP' implies in this case that the tasks need to be carried out by the printing firm. This could be avoided through the use of a different process, which would also enable a higher quality to be attainable. Essentially, any printing process can be used with DTP.*

Various printing processes

In general, it may be assumed that DTP publications will be produced by offset printing. However, in the following section we would also like to introduce briefly other current printing processes.

Planographic printing method, offset

Offset printing, small offset printing and lithography belong to planographic printing methods. Lithography was invented by Alois Senefelder in the 18th century and was superseded by offset printing at the beginning of this century.

With offset printing both the areas which are to be printed and those which are not lie in a single plane. The areas to be printed are separated from those which are not by the opposing physical–chemical reaction of certain substances to water. This means that areas which are to be printed repel water, i.e. they attract oil and hence colour, whilst the areas which remain unaffected are water absorbent (hydrophilic).

As mentioned already, various types of printing plates (aluminium, compound metal, plastic plates) with a light-sensitive surface, on to which camera-ready copy (film) is photoset, may be used for the offset process. These plates are mounted on to a cylinder. Before ink can be added the plates need to be wetted uniformly. The printer refers here to a 'wetting medium'. The wetting is achieved by means of so-called damping rollers. The areas which are to be printed do not absorb any water. Ink is then added to the plate cylinders by means of ink rollers, whereby the areas which did not absorb any water when dampened now absorb colour. Colour is first transferred to a cylinder covered with a rubber cloth and from there it is transferred to the paper (indirect printing).

Ever increasing printing speeds and larger printing formats are placing high demands on the paper used in offset printing. Important criteria here are stability under tension and the resistance to dust. Fibres or dust settling on the rubber cloth have a negative effect on the quality of the printing. Overall, however, offset printing is relatively flexible regarding different types of printing paper. Owing to the elasticity of the rubber cloth, even very rough paper such as cardboard, and paper with embossing or a texture, may be printed. In operation, offset printing may be categorised into sheet offset and roll offset. In sheet offset cut sheets are printed, whilst in roll offset a continuous paper strip is printed from a roll.

Offset printing is a very economical printing process. High printing speeds and good quality are attainable.

Letterpress printing, relief printing

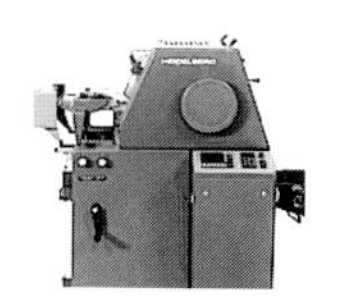

In letterpress printing, the areas to be printed are raised relative to those which are not. The areas raised to the same height accept ink when the printing form is inked and transfer it to the printing material. Included in letterpress printing is flexography and linear printing, as well as letterset (indirect letterpress printing).

Since letterpress printing has critical disadvantages when compared with offset printing, it is becoming less and less important. Even books are now produced using offset printing. Among others, the reasons for this are that the use of filmsetting is only possible under certain conditions, printing media consist of many individual parts, the printing preparation takes a relatively long time and even the printing performance is worse than that for offset printing.

Gravure printing

All the printing processes in which the areas to be printed are sunk into the surface of the printing medium are known as gravure printing. The picture areas to be printed are reproduced as tiny indentations in the printing medium by means of pricking, engraving, washing out or etching. Next the printing medium is covered liberally with ink and the excess 'wiped away' again. Under high pressure, the ink is 'sucked' from the paper to be printed and then from the printing medium. The most important applications are for magazines, catalogues, brochures and packing materials. As the complete printing medium cylinder needs to be processed when creating the printing medium, this process is only worth while for large print runs (about 500 000).

Screen printing

In screen printing the areas to be printed in the printing medium absorb ink. The ink is painted with a rubber squeegee over the screen and passes through the open areas on to the material to be printed. In screen printing, it is possible to use more ink and even extraordinary printing materials such as glass, metal, plastic, textiles, etc., may be printed on. It is used for simple linear printing pictures such as forms, areas and large texts.

Paper

Paper is manufactured from fibres mechanically or chemically set. The fibres usually originate from plants and are matted together when suspended in a water solution. Using processes of various complexity incorporating several additives, the most diverse types and qualities of paper may be obtained. Paper or cardboard is assessed

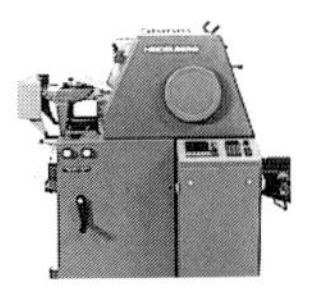

according to the following: material composition, surface finish, transparency, adhesiveness, weight, strength, area weight/specific volume, running direction (machine direction/direction of fibres in paper) and width.

Paper containing wood represents the largest group and is known as medium fine paper. This type of paper contains varying amounts of wood fibre (more than 5% by weight) and different quantities of bleached and unbleached raw materials as well as cellulose. The strength and the whiteness of paper decrease as the quantity of wood in it increases. Wood-free paper contains a maximum of 5% by weight of wood fibres. It is produced from varying quantities of rags and cellulose.

As it has already been explained in detail that the raster width is influenced by the paper quality, we would like in the following to describe the various paper surfaces.

- Machine smooth paper

  When the paper leaves the machine, it possesses a raw, matt surface, which is known as 'machine smooth'. In general this paper is used for the printing of newspapers, i.e. for a raster width of 75 lpi (30 L/cm) to 100 lpi (40 L/cm).

- Satined paper

  In order to attain a smooth and more shiny surface, the machine smooth paper is smoothed once or twice more in a so-called calender. This paper is also known as illustration printing paper or natural art printing paper and may be printed on using medium fine raster widths of up to about 140 lpi (54 L/cm).

- Coated paper

  The highest-quality surface is attained by coating the paper. In this case, filling and adhesive agents are added to the paper. This paper may be categorised into double-sided and single-sided coated paper and into gloss and matt coated paper. Coated paper may be printed on using very fine raster widths (300 lpi, 120 L/cm). Depending on the type it may also be known as picture printing, art printing and original art printing paper, where the quality and name depends on the type of coating mass, type and amount.

  Furthermore, other special surface processes also exist, such as embossing and compression.

Depending on the application various paper types with different weights (by area) are used. We do not, however, wish to describe them here. But before you decide on the paper for your specific application, you should consult the printers.

Chapter 10
Processing greyscale pictures

Anyone who thinks that they can ignore the subject of image processing when dealing with greyscale pictures is very much mistaken. Image processing programs are not only required to produce special picture effects or to organise picture layout, but above all they are also required for attaining optimum picture quality. Tone changes which are produced when photosetting and printing need primarily to be corrected. This may be carried out using a calibration or image processing program. The major problem in the processing and representation of photographs is that the final quality may only be determined once printing has been carried out. As already discussed, a printout from a laser printer can be ruled out immediately. Here, only the layout and not the quality of the photographs may be assessed. The pictures may best be assessed on the screen, provided it is capable of representing grey levels.

Processing images on a monitor which cannot display grey levels is completely worthless. Anyone who views a picture on a greyscale monitor for the first time will be impressed. Why is the greyscale representation better? Normally a picture

Without greyscale representation: no optimum image processing.

displayed on a monitor is rasterised or dithered. The individual grey levels are attained in the same manner as printing by means of various arrangements of dots. In order to represent 64 grey levels, 8 × 8 pixels are required for the rasterised representation of a single grey dot, as the brightness of a pixel cannot of course be changed. Owing to the rather low screen resolution of about 75 dpi it is therefore not possible to assess the screen picture.

However, a greyscale monitor requires only a single pixel instead of the 8 × 8 pixels, since in this case it may assume various levels of brightness. Consequently, in contrast to the laser printer, real grey levels may be represented. Therefore, in spite of the low monitor resolution of about 75 dpi it is possible to produce a picture of outstanding quality. Incidentally, it is not the monitor type which determines whether or not grey levels may be displayed, but the capability of the graphics card and software. A suitable monitor with graphics card is relatively inexpensive. Even if standard VGA monitors with VGA graphics cards are used, 64 grey levels may be represented, albeit with a lower resolution. If special VGA graphics cards with 512 kbyte memory are used 64 grey levels may be displayed with a resolution of more than 640 × 480 dots. If a program supports these features, it is possible to attain outstanding quality when displaying pictures with this system, so that the picture may be set and assessed to an optimum degree. Using standard graphics cards 256 colours may be displayed at the same time, but only 64 grey levels can be shown. However, 64 grey levels are sufficient for judging a picture.

Using an image processing program all the image parameters may be set and the result viewed immediately on the screen as an unrasterised greyscale picture. In addition, various manipulations of the picture may be carried out. Filter functions, such as sharpening and smoothing, are available as well as retouching. It is even possible subsequently to increase the image resolution by interpolating the grey levels. As the future lies with image processing, more and more software manufacturers will support the actual representation of grey levels. However, for each individual case you must enquire as to whether the selected hardware and software is capable of representing 64 grey levels, as well as a resolution of at least 640 × 400 dots.

This book was created on a 20" double-page monitor which could not display grey levels. For the image processing we switched to a black and white VGA monitor with greyscale representation. Black and white monitors are, incidentally, often better suited to judging single-colour pictures than normal VGA colour monitors. If a double monitor system is used, it is possible to work at the same time with a high-quality, ergonomic and inexpensive 20" monitor for the page layout, using a DTP program, and an inexpensive VGA monitor for the image processing. If possible, however, you should work with a 20" monitor which is also capable of showing grey levels, since then you do not need to switch continually between the two monitors.

With the aids and settings provided in this book, you will be able to obtain very good-quality pictures easily. However, the subtleties of image processing will remain a mystery to you if your system cannot display grey levels. Only when your work has been photoset will you be able to assess the result of your scanning, whereas with a greyscale monitor the scanner or image settings may be assessed immediately. If your scanner software does not enable grey levels to be displayed or displays them only partially, then it is recommended that an image processing program capable of showing grey levels is used.

Important terms

Before we describe the individual elements of image processing in detail, a few more terms are required.

Peak light, bright light, medium tone, shade

The following terms are used for the various grey or colour tones:

- Bright light or peak light for the light areas on a picture
- Medium tones for the medium areas on a picture
- Shade for the dark areas on a picture

Brightness/intensity

Grey levels are changed linearly with the brightness. If, for example, the brightness is increased, the picture becomes uniformly brighter. If the brightness is increased too much, the black areas become grey and the light grey areas white. With a large reduction in brightness dark areas become black and white areas grey.

Contrast

The number of grey levels is changed linearly with the contrast, i.e. fewer grey levels are printed. If the contrast is changed the medium grey tone remains unchanged. If the contrast is increased, light areas in the picture become even lighter and dark areas darker. In the extreme case only black and white areas are left. The picture at first looks as if it has more contrast, but then it approximates more and more to a line drawing.

Gradation (curvature, gamma curve)

The terms gradation, curvature and gamma curve all refer to the same phenomenon in image processing. The various grey levels may be changed by different amounts with gradation. For example, medium grey levels may be made brighter than very light or very dark areas on a picture. Unlike contrast and brightness, the complete picture is not uniformly changed with gradation; instead only selected individual greyscale areas are changed. With gradation the brightness and contrast are both set at the same time with the advantage that their values may be changed for each greyscale area. However, gradation may only be sensibly implemented if it is possible to enter both input and output values and if at least five support points are available. The input value is the grey level which the scanner or image processing program reads. If, for example, the scanner reads a grey area with a tone of 30%, then this value is the input value. The output value is the one which is to be saved instead of the input value. If, for example, the scanned value is to be made brighter by 5%, then the value set for output is simply 5% lower. In this example, the output value would then be 25%.

 *By setting the gradation a set of data may be read (input), changed (output) and then stored.*

The input and output values may be set for various grey levels. The more grey levels which can be defined (number of support points), the more precisely the gradation curve may be set. The values lying between the individual support points are calculated automatically. If the gradation can only be handled manually in an image processing program and the input and output values cannot be entered as numbers, then this function is more suitable for leisure purposes than serious work. In this case, using gradation it is certainly not possible to take into account the tone increase which occurs in printing. Using gradation it is possible, for example, to improve an original picture which lacks contrast, with the very light and very dark areas remaining almost unchanged. In addition, special effects may also be attained. If the contrast is reduced, light areas become darker and dark areas lighter. Then very light or very dark areas no longer exist. In such a case, the picture appears as if a grey veil has been laid over it.

Correction of tone increase

We will briefly repeat ourselves again by mentioning the problems which occur in photosetting, since even here the results may be distorted. If the paper or film is too soft and the photosetting process has not been calibrated, good results may be

obtained for line drawings and text, but reproducing photographs to a uniformly good standard can become a matter of chance. In order to obtain good pictures, you need to find a photosetting studio which calibrates the photosetter regularly and which also continually checks the tone increase. If photosetting is carried out on to standard film material (line film), the corresponding tone increase needs to be taken into account. In order to determine the exact tone increase with which the photosetting studio works, a test file including precisely defined grey levels may be photoset and the grey levels subsequently measured using a densitometer.

In general, however, if the values given in this book are used then sufficiently good results will be obtained. If all obstacles have been negotiated successfully and the photoset film is ready, then printing may commence. If errors occur during image processing or photosetting, it is possible, if necessary, to carry out the work again. However, this is not the case with printing. If the van from the printers has already arrived with the finished brochures or catalogues, then it is no longer possible to send it back. It is then a bitter experience if, having expended so much effort beforehand and paid considerable charges for printing, you are disappointed with the quality of the pictures. As mentioned before, as pictures become darker when printed, an important issue in image processing is the tone increase. The data is 'distorted' beforehand so that the end result corresponds to the original picture.

Why are pictures often too flat or too dark?

Anyone who has printed scanned pictures will probably have wondered why the pictures are so flat or so dark. Well, this is because the tone increase is not the same for all grey levels when printing. Medium grey tones increase in tone more than the light and very dark areas in pictures. If you try to set the picture to an optimum degree using the brightness function, both the medium grey tones and the very dark areas will become lighter. In the end the picture will lack depth and appear as if a grey veil has been laid over it. If, however, the brightness is not increased so much, the complete picture will become too dark. Hence, in order to obtain a consistently good picture each grey level needs to be corrected by a different amount. The very light and the very dark areas in a picture should remain unchanged. It is mainly the medium grey tones which are corrected.

Brightness and contrast need only be set in a few cases, since it is primarily the gradation which needs to be manipulated to set each grey level individually.

Carrying out tone corrections

In order to correct picture distortions which occur during printing and possibly photosetting, each grey level needs to be changed individually. This individual setting is, however, only possible if a so-called gradation can be set. By using a

gradation curve it is possible to determine by how much the individual grey levels should be changed. As stated earlier, the gradation curve is sometimes also known as 'curvature' or the 'gamma curve'.

Let us consider an original picture which has a grey level of 40% in a particular area. As a result of the tone increase during printing the grey level may rise to 55%. For this reason, the 40% grey level must be made lighter beforehand, so that after printing the area has a grey level of 40% once more. Often, however, the gradation may not be set at its optimum value in the scanner software and so the means of setting it there may not be used. It may perhaps be sufficient for the output from a laser printer, but we prefer to achieve rather more. If the scanner software is not capable of meeting the requirements, then an alternative is to use a suitable image processing or calibration program.

The exact correction of individual sources of errors is only possible if both input and output values can be entered to define the gradation curve.

The input value is the grey level which the scanner reads or the value which is stored in the data set. The output value is the new value which is to be saved instead of the input value. Below is an example showing how a gradation curve may be created using input and output values:

Input	Output
0%	3%
30%	17%
60%	39%
80%	58%
90%	76%
100%	95%

In the above example, an input of 60% and an output of 39% means that all 60% values are reduced to 39%. In addition, the 80% value is reduced to 58% and the 90% value to 76%. For printing reasons, which will be explained in the next paragraph, the value of 0% is changed to 3% and 100% is changed to 95%. All the intermediate values are automatically interpolated. In this way it is therefore possible to adjust the individual grey levels by different amounts. As exact numbers may be entered, the tone increase may also be taken precisely into account.

In offset printing the range of tones which may be reproduced lies between 3% and 97% (with respect to the film) at a raster width of 150 lpi (60 L/cm) for high-quality paper (coated paper). For this reason, when processing pictures care should be taken to ensure that near the edges no tone is under 5%, otherwise the border of the

picture will be missing. In addition, the tones should be less than 97%, since due to the tone increase all values over 97% will become uniformly black. For larger raster widths and poorer-quality paper the range of reproducible tones is reduced accordingly.

Rule of thumb: At the edge of a picture, the lightest area should have a tone of 5 %, otherwise the border of the picture will be missing.

If a picture is to exhibit considerable contrast, then the range of tones may also stretch from 0% to 100%. However, this depends on the type of picture and also on whether there are any very light areas near the edge of the picture. The individual values for correcting tone increases in printing are shown in Chapter 12.

Scanner calibration

What does calibration mean? Consider a record player which produces tones from the loudspeaker that are different to those on the record. You can adjust the bass and treble, but the tone will never be correct. In order to correct the wrong tones that this record player is producing, an amplifier is required which is capable of correcting each individual tone. If the tone 'C' is played, then the amplifier is tuned continuously until a tone 'C' is produced from the loudspeaker. This adjustment is carried out for all tones, so that after the adjustments are complete each recorded tone is output by the loudspeaker without distortion. Such adjustment is known as calibration. Of course, this calibration is unnecessary for record players, as it has already been carried out by the manufacturers.

As the grey levels which the scanner saves may deviate from the original, the scanner should be tested and its linearity checked. Linearity implies uniformity and in this case means the following: if an original picture is scanned on which the tone increases gradually and uniformly, then the scanner should likewise output grey levels uniformly from 0% to 100%. Usually, the values read by the scanner only deviate slightly from the original ones if scanning is carried out with the brightness and contrast set to 50%. Gradation should not be changed here; instead its curve should be a straight line at 45°. Sometimes 0% also needs to be entered for these medium scanner settings. If a value cannot be entered for the brightness and contrast for scanning, these values are used automatically.

The simplest way of checking a scanner is to scan black and white areas and then measure them using an image processing program. The white area should be approximately 0% and the black area approximately 100%. If the measured value

for the white area deviates only slightly from the control value (about 5% to 10%), then this not serious. In any case, different paper varies its whiteness. It is more important that the black value is approximately correct (deviation of about 5%).

In order to obtain a feel for the individual grey values and to check the scanner in more detail, it is recommended that a complete grey-scale wedge is measured. This is also instructive since scanners exist which already have a particular gradation included internally. A grey-scale wedge consists of a sequence of individual grey areas which increase in density by a precisely defined amount. The grey-scale wedge is scanned and then measured using the image processing program. In order to compare the control values with the measured values, it is necessary to know how the grey-scale wedge is defined. The problem is that it is usually defined logarithmically. As our eyes also see approximately logarithmically, we perceive that the steps in value are relatively equal.

If the grey levels are now printed as percentages, we see that the steps are not linear. The grey-scale wedge on the right shows that step 3 already corresponds to a 50% value. The darker the grey fields become, the less the percentage increase in the grey value. The first time anyone tries to assess a grey tone, they will probably believe that the 80% grey tone is the 50% one. You will be amazed at how light a

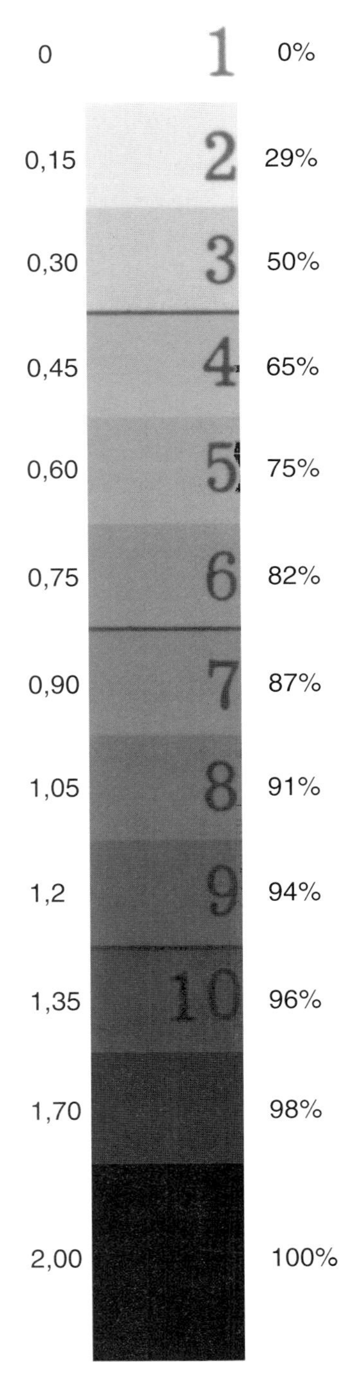

The picture on the right shows a 12 step grey-scale wedge. The density (logarithmic unit) is shown on the left and the grey value (percentage) on the right. The grey-level scale is defined in such a way that the increase in density from one grey area to the next is log 0.15. Please note that step 3 already has a grey tone of 50%.

50% grey tone is. With such a tone, 50% of light is reflected. With a rasterised 50% grey level, 50% of the area is black. The individual grey levels in the grey level scale are given as 'densities'. As density has already been explained, we will say only this here: density is the ratio between light shone on to and light reflected from the grey area and is given as a logarithmic unit.

If the structure of the grey-scale wedge is now clear, you may commence to check the scanner. The grey-scale wedge is scanned, with both brightness and contrast kept at a medium level (both brightness and contrast at 50% or 0% depending on the scanner). If it is possible to set the gradation, it should not be changed, i.e. its curve should be set to a straight line of 45°. Afterwards the grey-scale wedge is measured using the densitometer function within the image processing program and the values obtained are recorded.

With respect to scanner calibration it is important to remember that the grey-scale wedge is structured logarithmically. Hence, for an 11 step grey-scale wedge you cannot assume that step 1 = 0%, step 2 = 10%, step 3 = 20%, etc. With a grey-scale wedge you need to know by what value the density increases from step to step. Afterwards, by using a table, the actual percentage values may be determined. Such a conversion table is shown in Chapter 12.

In general it is better not to carry out scanner calibration. Instead, when processing the picture, it is preferable to take into account the individual scanner characteristics, especially when you have a scanner which includes a particular gradation curve set internally (e.g. due to a logarithmic amplifier). However, we have described this topic in detail in order for you to understand what calibration means and also because scanner calibration is sometimes described incorrectly.

Screen calibration

Next we will consider a very important point, often not given sufficient attention. The screen must be calibrated! Here we would like to describe a simple method, through which realistic a representation as possible may be obtained. It may not be 100% accurate but it is completely sufficient in practice.

If you do not carry out any special picture adjustments and wish only to consider the corrections for the tone increase (photosetting, printing), then it is not a problem if the screen displays the grey levels incorrectly. However, if you wish to set up a photograph to an optimum degree using the representation on the screen, then the

screen must be able to reproduce the photograph and its brightness values exactly. In contrast to the scanner, which outputs the grey levels reliably, the screen displays the picture in a completely different way to what it looks like in reality. For example, a 30% grey tone is displayed as a 50% level and a 50% level as an 80% grey tone.

 *A monitor displays a picture much darker than it is in reality.*

If you wish to adjust a picture on the monitor, you must be aware that the representation on the monitor has little in common with reality. Only when the sources of errors are known exactly, may fundamental image processing be carried out. Therefore, one of the most important tasks after installing an image processing program is to calibrate the monitor, so that the picture is displayed true to the original.

If you do not take into account all the problems from tone increase to monitor calibration, then you will only obtain good-quality pictures after many trials and much experience. A picture is adjusted, photoset and printed. As the monitor reproduces the picture too dark, the picture is made lighter and therefore, without realising it, the tone increase which normally occurs in printing has already been partly accounted for. If after printing, however, the picture is still too dark, the complete procedure may be carried out again. On one hand this method of working is expensive but on the other comprehensive experience is required in order to interpret the screen representation correctly. If, in spite of many trials, the pictures produced are too dark or too flat, or if changes in quality continually occur then you may end up with a mountain of unsolved puzzles.

This book should help you to reach your goal by the quickest method, i.e. a good-quality picture should not have to be obtained by means of trial and error, but quickly by using exact settings. If pictures are to be optimised on the monitor it is extremely important that the monitor can display them as faithfully as possible. However, the calibration of the monitor is not that easy, since it cannot be calibrated simply by adjusting the controls for brightness and contrast. Therefore, we will need to discuss this topic rather more deeply.

A grey level must be displayed as brightly on the screen as the scanner produced it. This is a problem in so far as the screen itself emits light, whereas for a paper picture light (e.g. daylight) needs to be shone on it and a colour or grey tone is produced due to the reflected light. Because of this a monitor can naturally never be set exactly, so that a displayed picture corresponds completely with a paper picture with respect to brightness. In addition, due to the lower resolution, as well as the difference in picture size, a slightly different impression may be created by a picture on a monitor. However, that is not so important.

The actual problem is that the screen displays grey levels distorted by different amounts. A grey level of 20% must be corrected by a different amount than a grey level of 80%. Therefore, it is also possible to say that tone changes occur on the screen. In the same way that the tone increase created during printing needs to be corrected, the distortion in tone produced on the screen needs to be corrected appropriately by the image processing program. However, a problem in correcting the monitor tone increase is that only the screen representation needs to be altered and not the picture data set. The data set of course contains the correct values, which are only being incorrectly represented on the monitor.

1. Basic procedure

In the following sections several methods of calibrating the monitor will be introduced. Usually it may be assumed that the data set is correct if scanning was carried out with the brightness and contrast set to 50% (or 0%), or if it was done using an image processing program. However, scanners exist which provide an internal gradation and so the data set is modified. Therefore, in all cases the scanner should first be checked. With colour photographs care must also be taken, as many scanners are not able to recognise all the colours to the same degree (drop-out colours). A different impression of brightness is created for a colour representation than for a black and white one. Therefore, a black and white original should definitely be used to calibrate the screen. As the scanner saves the correct values in the data set, the monitor must be set up so that the picture corresponds as closely as possible to the original.

⇨ *In order to calibrate a monitor, both a black and white photograph and a grey-scale wedge are laid on the scanner glass plate and scanned together. The grey scale and the photograph may then be assessed on the screen at the same time.*

⇨ *Next, the monitor and the gradation of the image processing program are adjusted so that the representation on the monitor corresponds exactly to the original picture (photograph and grey-scale wedge).*

⇨ *A colour photograph must not be used to calibrate a monitor.*

In order to adjust the screen the photograph is best placed directly adjacent to it. During adjustment individual parts of the picture should always be evaluated and the individual grey levels modified in order. Above all, the medium and dark areas of the

picture must be taken into account. The best representation is obtained using a monochrome VGA monitor. However, it is also possible to view the result using a colour monitor.

Monitors should always be set using the same photographs, so that values for both the monitor representation and the printed result may be obtained from experience. Consequently, the monitor should always then be set up consistently. Keep a few photographs as well as a grey-scale wedge available and also keep the scanned sample pictures. A good sample picture can also be created by scanning, for example, a picture composed of half a light photograph and half a dark photograph and the grey-scale wedges pasted together on a piece of paper. In this way different compositions may be observed at the same time on the screen. Large black and white photographs should be used for this purpose.

As mentioned already, adjusting the monitor using the brightness and contrast controls is not sufficient. The setting of the gradation in the image processing program will now be discussed. However, first the brightness and contrast controls are adjusted until at least the light tones are approximately correct. For this purpose the controls will often need to be adjusted until they can almost be adjusted no further.

2. Calibration function of the image processing program

The image processing program should include several possible applications for gradation curves (tone changes). The most common is where the data set for a picture (TIFF file) may subsequently be changed using a gradation curve. Furthermore, gradation or calibration curves may be used for the scanner or printer. In addition, it should be possible to set a gradation for the monitor, i.e it should be possible to calibrate the monitor. Here, only the representation is compensated for on the monitor. The data set for the picture remains unchanged. If the monitor is calibrated, the data set is displayed on the screen exactly as the scanner stored it. The picture may now be changed as desired, but you will always view it as stored in the file. Finally, only the tone increase which occurs in photosetting and printing remains to be taken into account.

Unfortunately, some software manufacturers do not understand, even after many discussions, that it is definitely necessary to calibrate the monitor. This paragraph is therefore primarily intended for software developers. Scan a 50% grey level. (Warning: This is the third and not the sixth step of a 12 step grey-scale wedge! If you do not believe it, ask a reproduction expert.) Therefore, scan the third step of a grey-scale wedge. Now it is possible, without the aid of a measuring device, to

recognise immediately that the grey level is far too dark on the monitor. That the grey level was correctly produced by the scanner may be checked easily using the densitometer function of the image processing program.

3. The simplest type of image processing

The simplest type of image processing is when nothing on the picture is changed manually and only the tone increase, which occurs during photosetting and printing, is taken into account. In this case the picture is loaded into the image processing program, a suitable gradation for the application (paper quality, raster width) is chosen and the picture is then saved again.

4. Calibrating the screen with the gradation

Not every program (yet) includes the capability of calibrating the screen. Hence, we have to find another means to do this. The gradation curve is used for calibrating the monitor since in any case it is required for the correction of the tone increase. This is possible because the tone increase in the monitor is similar to the tone increase in printing. As mentioned already, the picture needs to be made lighter for printing. For this purpose a gradation curve is created in the image processing program, whereby the corrections for the relevant tone increase are entered as input and output values. This gradation curve (set up for printing) is then loaded. For example, the required corrections for a tone increase of 22% (in the 40% field) may be used:

Input	Output
0%	3%
30%	17%
60%	39%
80%	58%
90%	76%
100%	95%

As the picture is made much lighter due to the gradation curve, it is then possible to adjust the monitor using the brightness and contrast controls so that the picture (grey-scale wedge and photograph) on the screen corresponds more or less with the original. Hence the original picture and monitor representation are identical.

As the scanned data set is correct and the gradation was only required for the adjustment of the monitor, the picture with the loaded gradation curve should not actually be saved. However, as a gradation curve is loaded, which is nevertheless required for the tone increase in printing, the picture may be processed and saved, thereby automatically taking the tone increase into account.

The simplest method of adjusting the monitor is through use of the gradation curve, which in any case needs to be used for the corresponding printing process. The values normally used for this gradation curve may be found in Chapter 12.

What does the correct image processing now look like?

The image processing following the monitor calibration is very simple. The picture is scanned and then set to an optimum degree on the monitor. The gradation, with which the monitor was calibrated, may be loaded as the basic setting. This gradation curve may then be changed as desired. If the picture on the monitor appears to be set to an optimum degree, it may then be saved. The gradation curve which was used to calibrate the monitor is automatically taken into account after saving and hence a particular tone increase is also already perceived. This implies that it is not the current or changed gradations which are important, but rather the gradation with which the monitor was calibrated in the first place!

A picture is adjusted visually on a monitor using the gradation. However, for the correction of the tone increase (printing) it is not this gradation curve which is important, but the gradation with which the monitor was calibrated in the first place.

For the purposes of clarification let us summarise:

- The data set from the scanner usually corresponds more or less to the original picture.
- The monitor is adjusted by means of a gradation curve, which is also needed for printing, in such a way that the representation on the monitor agrees with the original picture. This is necessary since standard monitors are not suitable for displaying the data set true to the original and because the image processing program does not possess its own monitor calibration.
- If the monitor is calibrated pictures may be scanned and adjusted, with the gradation curve changed as desired until the pictures are acceptable.
- After saving, the gradation curve for tone correction when printing has already been taken into account when the monitor was first adjusted.

Naturally, you should understand that by using this method an approximation only is attained. However, according to our experience, it has proved to be successful in practice.

5. Tasks with various monitor settings

As described already, the gradation for the monitor setting may be selected in such a way that it corresponds to the tone corrections required for printing, i.e. the tone increase produced in printing is already taken into account when calibrating the monitor. Once the monitor has been calibrated, pictures may be scanned and set to an optimum degree using the gradation on the screen.

This process, however, is tied to a particular tone increase and hence to a specific film material for photosetting and a specific quality of paper. If you wish to print on paper of various qualities or work with different raster widths, then a different tone increase needs to be taken into account each time. In order to use various tone changes, the monitor can be recalibrated each time to the actual tone increase. As before, a black and white test picture is scanned and loaded into the image processing program together with the previously scanned test picture. Afterwards, the desired gradation is loaded for the relevant tone increase. The monitor may now be set up so that the representation of the new picture and the original are the same.

Once the monitor is calibrated for a new tone increase, the actual picture may be scanned and the picture settings changed accordingly. The gradation curve may be changed as desired until the picture on the monitor is acceptable. Then it is saved. As the monitor was calibrated for a particular tone increase, this tone increase is saved in the picture data set. It is therefore completely irrelevant what the gradation curve looks like at the time of saving. This process, however, does have the disadvantage that the monitor needs to be adjusted continually. A method for using different gradation curves is described in the next section. However, you should realise that this somewhat time-consuming task is only necessary if your image processing program does not have the facility of monitor calibration.

You should only apply the following section if you are already familiar with the previous methods.

6. Working with various gradations

If you wish to retain flexibility and do not wish to reset the monitor each time, the actual tone increase may be taken into account as follows. A gradation is used for monitor calibration with which the monitor can be adjusted so that it is just light enough. In the best case this would be the gradation which was created for photosetting on to film with lith characteristic and for printing on to high-quality paper (art printing paper), i.e. this is the gradation for the lowest tone increase. Suitable values may be found in Chapter 12.

If the picture is then processed and saved, only a low tone increase has been taken into account. If, however, paper of a lower quality is now used, then a larger tone increase occurs than has been incorporated by the monitor calibration. Therefore, after the picture has been adjusted and saved, the difference between the perceived tone increase and the required tone increase for the data set needs to be accommodated. If, for example, the monitor is calibrated with a 19% tone increase, but an actual tone increase of 25% occurs, then the incorporated tone increase is 6% too low.

It is, however, not that simple to use the difference between the individual gradation curves to change the tone increase again. Example: An 80% value is to be changed to 62% for art printing paper and to 54% for paper of a lower quality. As the tone increase for the art printing paper has already be taken into account (by means of the monitor calibration), only the difference between the art printing paper and the paper of lower quality should now need to be reconciled. However, as the 80% value was first changed to 62%, in the second change it is not the former which needs to be changed but the latter. In practice this means that for each further change in gradation the previous output value must be used as the input value.

Let us have a look at concrete values for the tone increase associated with individual paper classes:

Input	Output Paper class 1	Paper class 2	Paper class 3
0%	3%	3%	3%
30%	18%	17%	15%
60%	41%	39%	36%
80%	62%	58%	54%
90%	81%	76%	71%
100%	98%	95%	94%

The 80% value is read by the scanner and represents the input. As a result of monitor calibration this value is changed, for example, to 62%. This corresponds at the same time to a tone correction for paper class 1. If, however, printing is to be carried out on paper class 3, in the second correction the 62% (and not the 80%) value needs to be changed to 54%. Here all the previous outputs are used as inputs and the new values for paper class 3 are entered as outputs.

The following example clarifies how the difference between paper class 1 and paper class 2 is taken into account when the tone increase for paper class 1 has already been saved:

New Input	Output Paper class 2
3%	3%
18%	17%
41%	39%
62%	58%
81%	76%
98%	95%

The gradation for the difference between paper class 1 and class 3 is as follows:

New Input	Output Paper class 2
3%	3%
18%	15%
41%	36%
62%	54%
81%	71%
98%	94%

In order to understand the relationships better, it is worth while examining the tables on tone increase and tone correction given in Chapter 12. If the principle has been understood, then the method is relatively easy to use. If a picture is loaded, adjusted and saved, then for further tone corrections the output of the monitor calibration needs to be used as the new input. The relevant values from the tables are entered as outputs as before.

In the previous examples we assumed that the monitor was calibrated with a gradation for paper class 1. However, if a second processing is subsequently carried out for the actual tone correction, then any values may be used for monitor calibration.

Suppose that your monitor is not adjusted to an optimum degree on the basis of the correction for printing. In this case, the gradation curve is adjusted until the representation on the monitor corresponds with that of the original document. This gradation curve may then be used to calibrate the monitor in future. The outputs used here should be noted for further processing. Afterwards, the picture may be set up as usual. At the end of the image processing and after saving, the desired tone increase still needs to be taken into account. This occurs when the output values from the monitor calibration are entered as inputs. If tone changes are reversed in doing this, then the quality of the picture will suffer. This may be necessary if the monitor

cannot be set to be light enough (particularly with colour monitors). Here a gradation will need to be used which is created for a large tone increase. The picture which was adjusted to be too light beforehand must be made darker again afterwards. Greyscale information disappears in this process. If image processing does not need to be carried out, it is better if the picture is not saved, but that the desired gradation is loaded first and then the picture saved.

An example for image processing: Once the picture has been scanned, the gradation curve with which the monitor was set up is loaded. The picture may then be processed and the gradation changed and assessed on the monitor. After image processing the picture is saved and hence the tone increase which was perceived when calibrating the monitor has already been taken into account. If paper of a lower quality is then to be used, a gradation curve is loaded or created, whereby the outputs of the monitor calibration are entered as input values. The picture is then saved once again. The main task consists in familiarising yourself with this subject and then creating and saving the relevant correction curves once. Image processing may then be carried out simply and quickly, since only the curves actually required still need to be loaded.

7. An example of monitor calibration

In order for you to understand this issue better, this section will provide a concrete example of how the monitor may be calibrated using the setting of the gradation and then how image processing is carried out.

Please note that the time-consuming monitor calibration mentioned here is only necessary because some software manufacturers have totally overlooked the issue and are still not yet convinced that monitors need to be calibrated.

Monitor calibration

- A black and white picture is scanned.
  (Scanner is set to 50% or 0% for brightness and contrast.)
- A gradation curve with the following values is loaded or created:

Input	Output (Paper class 2)
0%	3%
30%	17%
60%	39%
80%	58%
90%	76%
100%	95%

- The monitor is now adjusted by turning the controls for brightness and contrast (and by the monitor gamma of the image processing program) until the picture on the screen is as close to the original as possible.

  The monitor was calibrated in this way so that the tone increase for paper class 2 was automatically taken into account. Put another way, in using this monitor setting you are not observing the representation of the scanned data set, but are viewing the picture as it will appear after printing at a particular tone increase.

Scanning and setting of pictures

- Once the monitor has been calibrated, the desired pictures may be scanned.
- After scanning, the picture is adjusted on the monitor to an optimum degree by means of the gradation. Here, it is best to load the gradation which was used for the monitor calibration. This gradation may now be adjusted as desired until an optimum representation of the picture has been attained. As few changes as possible should be carried out since the monitor does not reproduce areas on the picture, especially dark ones, exactly.
- The picture is stored after the picture adjustment. The tone increase which occurs in printing has therefore already been taken into account by the calibration of the monitor. The picture may be linked to a DTP document without further processing and then sent to be photoset.

Paths to success

Actually, it must now be clear to everyone what the path to success looks like. In order to obtain optimum pictures, you must recognise the individual errors, which may occur during scanning, image processing and printing. Only when these errors are understood fully can the picture data set be suitably modified. Therefore, image processing primarily involves taking into account the tone changes which occur at the output device. Only by taking into account these tone changes which occur in photosetting and printing can a good-quality picture be produced. For this reason also during scanning a medium setting (50% or 0%) should always be used for brightness and contrast and all picture adjustments should mainly be made by means of the gradation. Changes in brightness and contrast should only be undertaken sparingly.

Values may be set for photosetting and printing which must be adhered to by everyone concerned. It is only when adjusting the screen that features sometimes still do not appear satisfactory. As the monitor distorts scanned pictures, the scanner or

image processing program should be able to calibrate the monitor. How this is achieved in individual cases depends on the capabilities of the program used and has already been described in detail. The monitor needs to be set so that the picture represented corresponds as closely as possible to the scanned original.

Printing using standardised values

Whenever mention is made of working with control values or to standards, it is implied that work is not to be carried out with the values that the photosetting studio or printing firm use more or less accidently, but that it is to be carried out using values which are generally accepted as standard values. This means that everyone concerned should adhere to the agreed values. Hence in photosetting, work should be carried out using film with litho characteristic and so only a negligible tone change should occur. The printing firm should work to the regulations of a standardisation, or they should agree contractually to a particular tone increase.

Before a picture is photoset, the tone increase which occurs in photosetting and printing needs to be taken into account and the data set for the picture (TIFF file) needs to be 'distorted' accordingly. For this purpose the values given for tone corrections in Chapter 12 may be used. These values are entered as input and output values in the image processing program and then saved. However, note that the current corrections are used and not the actual values for the tone increase. Do not worry if the picture appears to be far too light on film, as it is only after printing that you will attain the desired result. Which corrections should be used for individual cases may be deduced from Chapter 12.

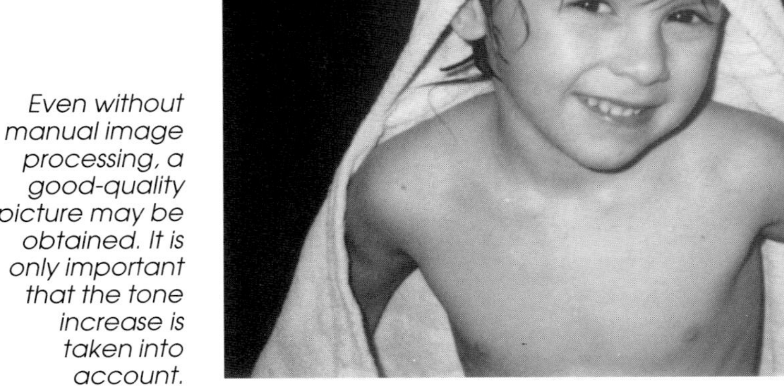

Even without manual image processing, a good-quality picture may be obtained. It is only important that the tone increase is taken into account.

A dark original picture must be made lighter manually using an image processing program ...

However, it is also sufficient if just a higher tone increase is set (by about 20%). Hence with only a few standard values a good-quality picture may always be attained.

Of course, if a manual adjustment is carried out, better results are obtained as the picture may be set more precisely and, for example, it may also be sharpened.

Determination of total errors

If your photosetting studio does not operate under the recommendations in this book, or if your printing firm cannot inform you of the exact tone increase used, which you wish to know precisely for your particular application, then it is worth while measuring the total tone changes which actually occur in practice.

For this purpose both a test data set and a scanned grey-scale wedge are photoset and printed. After printing, the grey levels may be measured using a densitometer. The values for this test data set provide the tone changes which occur, but exclude the distortions which occur in scanning, whilst the scanned grey-scale wedge exposes all the errors.

Essentially, you must be aware that each tone increase implies a decrease in quality, which may not be completely compensated for. So, for example, if distortions of the picture already occur in photosetting, then these will not be completely compensated for by carrying out a correction, since the range of grey levels (number of grey levels) is reduced continually.

Working with an image processing program

The advantage of an image processing program is that the picture can be viewed on the screen and manipulated if necessary. If the picture is set correctly, then the correction for the tone increase may be carried out. However, this can only be achieved with an image processing program which includes the facility of setting gradation curves using input and output values. It is important when working with an image processing program that the monitor may be calibrated and that the representation on the monitor corresponds as closely as possible to the scanned picture.

Working with a calibration program

The tone increase may also be compensated for when using a calibration program. Two types of calibration program exist. Compensation is carried out either in the TIFF picture file or in the PostScript printing file.

Calibration programs exist which provide the photosetter with instructions describing which tone changes need to be carried out. These instructions are written into the PostScript printing file and are valid for all the pictures found in this file. The advantage of such a calibration program is that the setting of the raster width and raster angle is carried out immediately and so cannot be forgotten. In addition, all the settings may be changed at any time or the original state restored.

Tips for optimum-quality pictures

Now that the most important relationships are clear, we would like to examine several practical applications. We would like to give you some tips on how to optimise the quality of pictures.

Interpolation (SmartSizing)

Previously it was stated that the required image resolution needs to be determined before scanning and that this value then needs to be set as the scanning resolution. However, if we desire optimum pictures, the difference between these two resolutions needs to be clear:

- ❐ The TIFF data set is defined using the image resolution which is specified as the number of grey levels which should be available for each raster dot to produce optimum output.

Left: *Scanned from a four-colour brochure at 100 dpi: moiré patterns and fuzzy lines were the result.*
Right: *Scanned at 300 dpi with the resolution later reduced to 100 dpi: moiré patterns can no longer been seen. The irregularities in the lines and edges have also been removed.*

In order to obtain the desired number of grey levels the corresponding scanning resolution (= image resolution) is calculated beforehand and then set.

❐ The scanning resolution is specified by the accuracy with which a scanner scans a picture. Many scanners operate on the following principle. For a scanner with a physical resolution of 400 dpi only every fourth sensor is activated if a picture is scanned at 100 dpi, i.e. the picture obtained possesses corresponding holes.

In greyscale pictures which no not contain any sharp edges or lines, this inaccuracy is hardly noticeable. However, in pictures which contain sudden transitions from light to dark areas, losses in quality may occur, especially if the scanning resolution is less than half the physical resolution of the scanner. This loss in quality is noticeable, in so far as lines which are straight on the original now appear stepped. This loss in quality may be reduced if scanning is always carried out at maximum resolution, with the required resolution then calculated internally by means of interpolation. For example, from four scanned grey levels a single average value may be saved in the data set. As many scanners are not capable of carrying out this interpolation, we need to perform it ourselves using a suitable image processing program.

Hence, whenever distinct straight edges appear in a picture and the calculated image resolution is less than half of the physical resolution of the scanner, we should take further steps to improve quality. Scanning should be carried out at either the maximum physical resolution, or at least half the physical resolution. Finally, using an image processing program the scanned data set should be reduced to the required size using the 'interpolation' function also known as 'SmartSizing'. As mentioned already, during interpolation the missing grey levels are recalculated and so individual grey levels are not just omitted. Once the data has been reduced by means of interpolating the grey levels, a data set then exists where each point of the original picture has been taken into account.

If you do not have the 'interpolation' or 'SmartSizing' function it is not worth while scanning at high resolution and then reducing the data. If a program does not offer data reduction by means of interpolation, then calculation of the average values may be carried out before data reduction at double the scanner resolution. In addition, smoothing may, if necessary, be carried out before the data reduction. In principle the same points apply as in Chapter 7, with the difference here that the missing scanner functions are less evident (no moiré effect, but untidy lines instead) and the picture needs only to be sharpened.

Before you match and calculate the resolution at which scanning should take place in each individual case, it is always simpler to scan at the full optical resolution and then to reduce the data set again afterwards. The desired size of picture and the image resolution are then entered for the data reduction. As already mentioned

several times in this book, when, as in this case, the image size is subsequently entered, the image resolution may be calculated as follows: Image resolution = raster width × 1.4 (or 1.2)

The best results may be obtained when scanning is carried out at a higher resolution (max. physical resolution of usually 300 or 400 dpi) and the data set is interpolated to the required image resolution afterwards.

If scanners which already carry out internal interpolation are used, then this point may be omitted. However, even when these scanners are used better results are often obtained using the above process, since the internal interpolation of the scanner is often only carried out for each horizontal line and in very rough steps.

The method of first scanning at high resolution and then carrying out data reduction has one disadvantage. A longer processing time is required since a very large file is created, which is then processed further and then finally reduced in size. However, the 'interpolation' function may also be used to increase the resolution. In this case new grey levels are calculated and added to those already present. Not only is the number of grey levels increased (as happens in some scanner software), but also a new grey level is calculated from the values of neighbouring grey levels. For example, the image resolution in video images may be increased via interpolation. Even small original pictures such as passport photographs or slides may be enlarged significantly using this method.

Setting of light and shade

The light areas on a picture are known as light and the dark areas as shade. In order to obtain a picture with contrast, very light and very dark areas should exist within it. It is sometimes better not to have a depth drawing (detail in dark areas) so that a picture with more contrast is obtained. Therefore, light and shade should be set immediately after scanning. In general, this means that the lightest area on the picture recorded by the scanner is set to 5%; the darkest area is set to 95%. This may be carried out either manually or automatically through one of many programs. So, for example, the darkest area recorded by the scanner may be changed from 83% to 95% and all the intermediate grey levels expanded accordingly.

In order to obtain very dark pictures readily, scanners exist through which the light intensity may be increased. The corrections which need to be carried out invariably depend mainly on the quality of the original and the composition. If, for example, the lightest area lies at the centre of the picture, it may be highlighted properly by reducing the tone to 0%. However, care needs to be taken here to ensure that the

Original scan

Light and shade have been set.

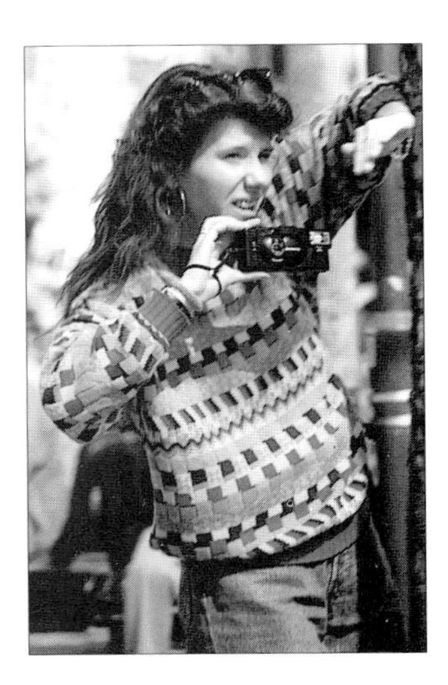

Standard gradation, sharpened

lightest area on the edge of the picture is at least 5%, otherwise the border will be missing. Having set the light and shade, the tone increase which occurs in photosetting and printing must be taken into account by setting a suitable gradation curve.

Tone comparison

Sometimes when adjusting pictures on a monitor you are not quite sure what a grey level will finally look like on paper. The grey-level region of interest may be measured using the densitometer function in an image processing program and the corresponding location on the grey-scale wedge may then be found. However, when this measurement is carried out the grey levels must not already have been made lighter for the tone increase. In addition, you should know whether the image processing program shows the original value saved in the data set or whether it shows the subsequent value already changed by gradation. If the value shown is the one saved in the data set, then the information must be saved before the measurement is carried out.

Sharpening

Through sharpening techniques the contours present in an image are made more prominent, so that details within a picture are highlighted. The 'video errors' of the scanner optics may be corrected using sharpening. When pictures are sharpened they appear to have more contrast and so appear more vivid. A better impression of the picture is therefore obtained and so, in general, pictures should be sharpened. The one exception is that of rasterised pictures.

Left: normal
Bottom left: sharpened once
Bottom right: sharpened twice
Dot shape: elliptical

Therefore, 'sharpening' is a filter function which processes a picture in such a way that not all grey levels are changed to the same degree, but instead the picture is changed according to the particular composition. With sharpening, the contrast of individual parts of the picture is increased. If the difference between the brightness of neighbouring grey levels is large, then this difference is amplified even more. Neighbouring grey levels which are very similar remain unchanged. This means that a fuzzy picture gains more contrast and becomes sharper. However, limitations do exist. If pictures are sharpened too much, zigzags and waves occur in lines. This is especially the case when these lines are very stark (e.g. black lines on a light background).

Right: sharpened three times
Bottom left: sharpened four times
Bottom right: sharpened five times
Dot shape: elliptical

In general, pictures should always be sharpened at least a little. An exception to this is pictures of rasterised originals. How much sharpening should be carried out depends on personal taste.

The choice of sharpening filter and the intensity of the sharpening depend primarily on the composition and application of the picture. In the final analysis, however, it is a case of personal taste. If a picture is sharpened several times, special effects may be obtained. Several examples of this will now be described.

Left: normal
Bottom left: sharpened once
Bottom right: sharpened twice
Dot shape: elliptical

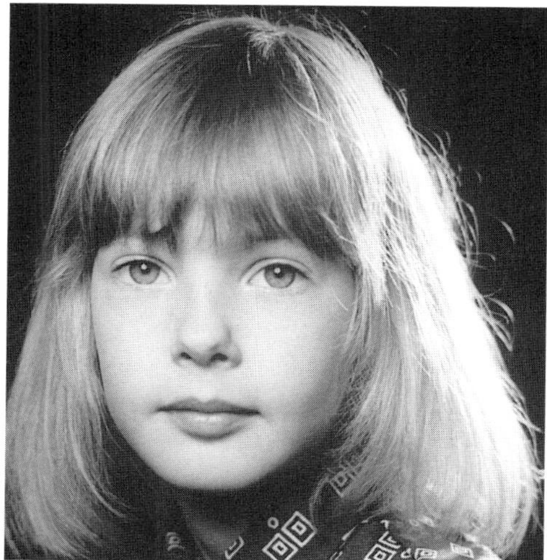

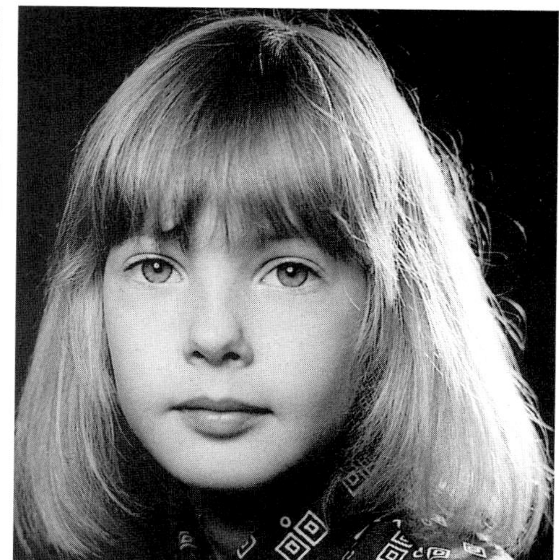

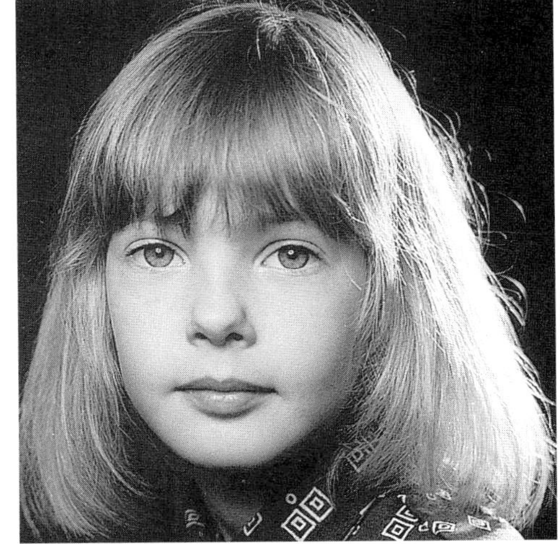

Right: sharpened three times
Bottom left: sharpened four times
Bottom right: sharpened five times
Dot shape: elliptical

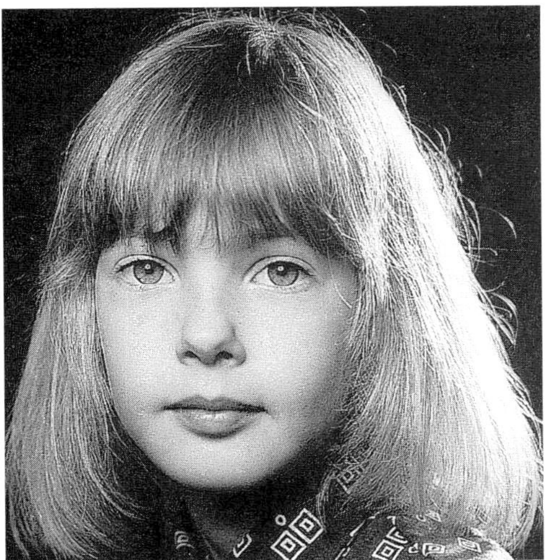

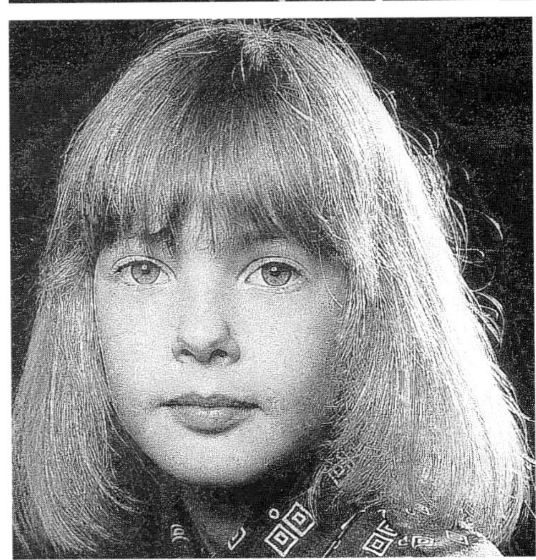

Top: normal
Centre: sharpened once
Bottom: sharpened twice
Dot shape: elliptical

Top: sharpened three times
Centre: sharpened four times
Bottom: sharpened five times
Dot shape: elliptical

Smoothing

Smoothing is also a filter function and produces the opposite effect to sharpening. Abrupt grey-level transitions are made softer. Smoothing is required, for example, when moiré effects, which are created through scanning rasterised pictures, need to be reduced. Functions also exist which may be used to smooth certain parts of the picture. For example, when parts of the picture are copied and moved the transition may be matched to the background.

Top: normal
Centre: smoothed once
Bottom: smoothed twice
Dot shape: elliptical

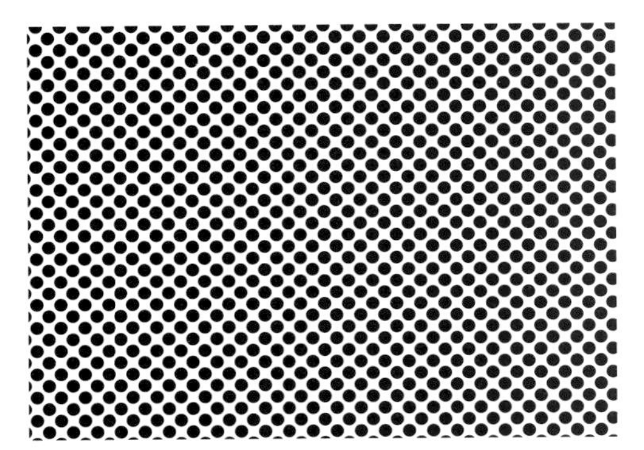

Circular dot

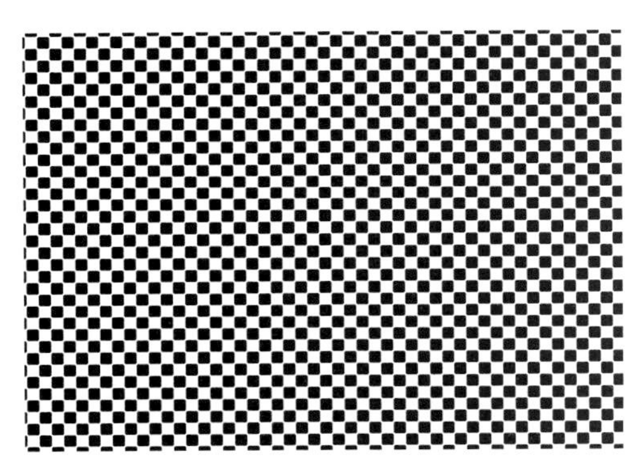

Square dot

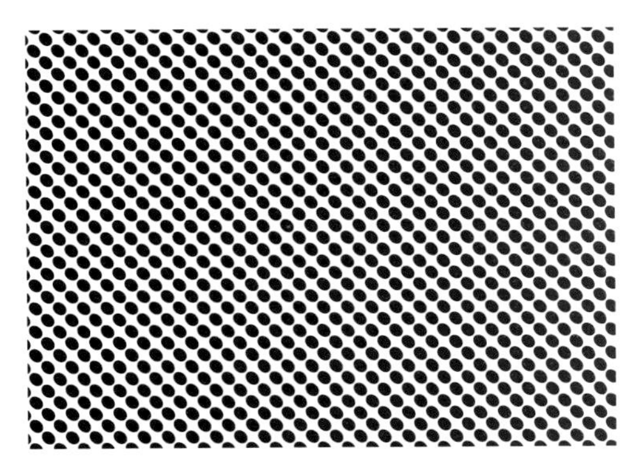

Elliptical dot

Dot shape

We have already mentioned that in single-colour prints the raster angle should always be 45°. However, until now the actual shape of the dot has been ignored.

A raster dot consists of a cluster of individual pixels. As a result of the different arrangements of the individual pixels various shapes of raster dot may be formed. Circular, square and elliptical dots are mainly used. Also, each shape may have several variations. For example, an elliptical dot may be more, or less, eccentric. Additionally, it is possible to have combinations of various shapes.

What is the point of having dots of various shapes? First, let us make sure that we understand how a grey level is simulated. As a raster dot is made larger the tone becomes darker. If the dot is increase continually at some point it will naturally collide with a neighbouring dot. This merging of individual dots is known as dot stop. At the moment the dots merge the grey level is increased abruptly. This is known as a tone jump. This tone jump is not noticeable in usual circumstances. However, where the grey levels change gradually this tone jump may be undesirably noticeable. This is above all the case when the dot stop occurs where grey levels change gradually over a large area and hence, in this tone range, a somewhat dark grey level suddenly appears.

If raster dots of different shape are used, the dot stop, and consequently the tone jump, may be influenced. The dot stop occurs at a grey level of about 40% for a square dot, at about 65% for a circular dot and at about both 50% and 75% for an elliptical dot. These values are naturally dependent on the exact shape of the dots as well as the tone increase. However, it is not only the dot shape which influences the tone jump, but also the raster width. The tone jump which occurs due to the dot stop is less noticeable in coarser rasters.

If square dots are used, the tone jump occurs in a lighter range than if circular dots are used. As mentioned already, in contrast to square and circular dots, with elliptical dots the dot stop occurs at two grey levels. This has the advantage that the tone jump

Scanned gradual progression of grey levels.

The characteristic dot shape may be recognised at a grey level of 50%.

Dot shape:
left: circular
centre: elliptical
right: square

is distributed and so is less distracting. In addition, the second tone jump for elliptical dots occurs at a relatively dark grey-level area and is therefore less noticeable. In contrast, tone jumps which occur in the lighter areas are more noticeable. It must also be mentioned that square dots are only square in the 50% grey-level area. In the remaining grey-level areas they take on a circular shape. Circular and square dots are therefore useful in circumstances where detail is of primary importance. If, for example, you wish the dot stop to occur at about 40%, because an even progression in the grey levels is required for the darker tones, then square dots may be used. There is no problem in printing square dots and this is usually set as standard.

Scanned gradual progression of grey levels at a raster width of 150 lpi.

Dot shape: left: circular centre: elliptical right: square

Elliptical dots are used for pictures with very fine grey-level progressions. However, as may be seen from the figures, the differences are not very great and so many reproduction experts use elliptical dots as standard. Another issue which should be mentioned regarding dot shapes is that they are not only present, and hence selectable, in the photosetter (RIP), but also may be created by the DTP program. The DTP program actually saves the formula for the current dot shape being used at the beginning of a PostScript file. This means that which dot shapes may be used and what they look like individually depends on the DTP program. However, care must be taken in creating the dot shape, since the description given in the DTP program is

Left: circular dot
Bottom right: elliptical dot
Bottom left: square dot

sometimes misleading. For example, when selecting 'dot' under dot shape a square dot is usually set. However, programs also exist which create circular dots when 'dot' is selected. If the standard setting of the DTP program is selected, then the current presetting of the output device is used. For a PostScript laser printer this may imply circular dots and for a photosetter square dots. The current dot shape may be easily checked. A grey-level progression is scanned, a raster width of about 10 lpi is set and this grey-level progression is then photoset. At this raster width the dot shape may be observed without the use of a magnifying glass.

Right: circular dot
Bottom left: elliptical dot
Bottom right: square dot

Besides the types of rasters already mentioned, so-called special-effect rasters also exist. These are rasters with special structures (straight, wave shaped or round lines) which are used to obtain graphical effects. Furthermore, so-called scatter rasters (grain rasters) exist, where it is the distance between individual pixels and not the dot shape or size which is changed. If a scatter raster is used pictures with considerable detail may be printed perfectly, but grey-level progressions, as present, for example, on a human face, are attained less successfully.

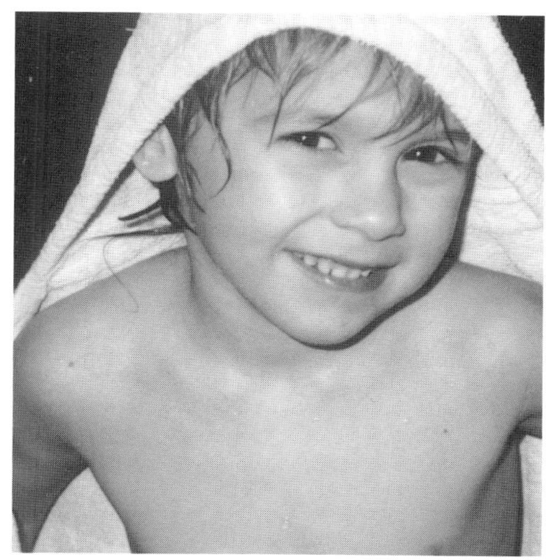

Left: circular dot
Bottom left: square dot
Bottom right: elliptical dot

Dot shapes:
Top: circular
Centre: square
Bottom: elliptical

Dot shapes:
Top: circular
Centre: square
Bottom: elliptical

Apart from the standard shape dots so-called line rasters, for example, exist which may be used to obtain special effects.
Top: 60 lpi (24 L/cm)
Centre: 120 lpi (47 L/cm)
Bottom: 150 lpi (60 L/cm)

We are processing a picture

We admit that many problems are highlighted in this book. If because of this the impression has been gained that the printing of pictures is extremely complex, then we must correct this impression before we go any further. By describing the problems which occur we wish primarily to make it easier for you to understand the basic relationships. In practice, however, good results may be attained without any deep understanding of the subject through means of suitable aids. In what follows we will describe the complete process 'from scanning to printing' to demonstrate how easily and quickly pictures of high quality may be integrated into DTP documents.

It goes without saying that image processing needs to be made even simpler in the future. Here we wish primarily to describe the facilities which are available on the market at present or which have at least been announced. Up till now we have tried to describe all the tasks required as generally as possible, so that the processing of images could be carried out with any suitable program. However, as many paths lead to the same final result, it sometimes appears more complex to achieve it than it is in reality. On the basis of the background knowledge gained here it is possible to develop a fairly practical way of working using the hardware and software available in a particular case. Perhaps through this book we may even encourage software developers to change their opinions and subsequently obtain programs which are more user-friendly and where the aspects of quality are given priority instead of the retouching functions. We now wish to show by using a suitable image processing program what the process to obtain a picture of optimum quality is in practice.

EXAMPLE 1:

Selecting section of picture

In the image processing program the 'scanning' menu is selected and a so-called 'pre-scan' is carried out to display the complete page on the screen. The section of the picture which is to be scanned is then fixed using a frame.

Scanning

Next, scanning takes place at the optical resolution of the scanner (usually 300 or 400 dpi). The image size remains at 100% for this purpose.
Remark: The process of scanning at a high resolution with subsequent data reduction may be ommitted when scanning photographs if HHR scanners (High Resolution Reduction) are used, i.e. if a scanner already performs an interpolation internally. However, the internal interpolation process of the scanner is often not good enough (interpolation is frequently only carried out line by line) to obtain a good result, even when using rasterised originals.

Original scan, file size: 1.4 MByte

After data reduction, file size: 0.35 MByte

Data reduction

The data is then reduced. For this purpose the required image resolution (raster width × 1.4) is read from a table, with respect to the desired image size for the DTP document, and then input. Once the interpolating function has been activated, the data is then reduced. If an interpolating function is not available the picture may be smoothed to a considerable degree before the data is reduced.

Storage

At this point the picture should definitely be saved, so that if the subsequent image settings are incorrect, the whole process does not need to be repeated.

Smoothing

(Only used for a rasterised image.) If a rasterised picture is scanned, but severe moiré effects can still be seen, then in exceptional circumstances it may be necessary to carry out smoothing using a suitable filter.In principle, smoothing may be carried out before data reduction.

Setting of ligth and shade

The lightest grey level is set to 5% and the darkest to 98%. In order to obtain pictures with high contrast the darkest grey level may be set to 100%. If the lightest grey level lies within the picture (not on the edge) it may be set to 0%.

Sharpening

(Not used for rasterised images.) The picture may next be processed using a medium or strong sharpening filter. If a rasterised original picture was scanned, then usually it should not be sharpened.

Taking the tone increase into account

Next, the relevant gradation curve is loaded to take the tone increase into account, which occurs on photosetting and printing. The monitor must be set up (calibrated) in such a way that the image corresponds as closely as possible to the original picture. The image must not be saved now.

Image optimisation

The gradation curve which has been loaded can now be changed according to personal taste and hence the picture may be optimised. However, a prerequisite for this is that the monitor should have been calibrated beforehand using a black and white picture (see previous point). Depths should only be made lighter by small degrees so that no tone outlines become visible. The problem which occurs when adjusting the picture is that although many graphics cards/monitors have 256 colours, they are only able to represent 64 grey levels. Therefore, in reality a drawing may have a certain depth which is not capable of being shown on the monitor. Only when the image has been optimised should it be saved.

Second Sharpening

In general, most pictures can be sharpened twice in order to obtain high-contrast images. However, you should be careful with pictures which contain very fine grey-level progressions, e.g. human faces. Otherwise the grey level progression could become too grained and so undesired details (like e.g. skin irregularities) may become too noticeable. In addition, at sharp edges zigzags may occur. So you should not sharpen, if the quality of detail in the picture is likely to suffer. Please note, when optimising a picture on the monitor, attention should be paid to the fact that the view on the monitor is adjusted to 100%. If it is less than 100%, lines could appear wavy or zigzagged on the monitor, which in reality is not the case.

Retouching

If it is desired to carry out an optimum retouching then it should be done before the data is reduced. However, it is naturally quicker and simpler to carry out retouching at this point.

Remark

The process described here may take a little longer as scanning is always carried out at a high resolution. However, all decisions regarding which resolution should be used for scanning and what type of image processing should be carried out are removed. If this process is used you can be sure that you will always obtain a picture with optimum quality. The 'interpolation' function ensures that the data set is optimised, regardless of whether the picture is to be reduced or enlarged. This example also shows that for standard applications it is better to have a 300 dpi

scanner rather than a 600 dpi one, since when the latter is used a much larger data set needs to be processed and reduced. This then takes correspondingly longer. It is only for line drawings or greyscale pictures which need to be considerably enlarged that 600 dpi scanners exhibit their advantages.

Top: In addition to the previous pictures bright and medium tones have been lightened slightly and dark tones stronger.

Bottom: Dark tones have been lightened even more.

EXAMPLE 2:

Original scan
Figure 2a

Light and shade have been set.
Figure 2b

Sharpened
Figure 2c

Standard gradation
Figure 2d

Background (wall) set to 0%.
Figure 2e

Sharpened a second time
Figure 2f

EXAMPLE 3:

Original scan
Figure 3a

Light and shade have been set.
Figure 3b

Sharpened
Figure 3c

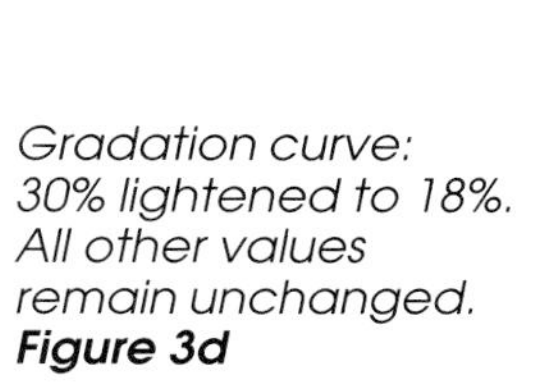

Gradation curve:
30% lightened to 18%.
All other values
remain unchanged.
Figure 3d

Gradation curve:
30% lightened to 18%,
60% to 41%.
All other values
remain unchanged.
Figure 3e

Gradation curve:
30% lightened to 18%,
60% to 41%,
80 % to 62%.
All other values
remain unchanged.
Figure 3f

*Gradation curve:
30% lightened to 18%,
60% to 41%,
80 % to 62%,
90% to 82%.*
Figure 3g

Sharpened a second time
Figure 3h

*Medium tones have been
lightened again.*
Figure 3i

The entire picture has also been lightened.
Figure 3j

Manually adjusted, faces of the persons manually smoothed, bottle and glasses sharpened.
Figure 3k

EXAMPLE 4:

Original scan
Figure 4a

Light and shade have been set.
Figure 4b

Sharpened
Figure 4c

Standard gradation
Figure 4d

In addition to the standard gradation bright and medium tones have been lightened by approx. 15% and dark tones by approx. 3%.
Figure 4e

In addition to the standard gradation bright and medium tones have been lightened by approx. 20% and dark tones by approx. 17%.
Figure 4f

Same as Fig. 4f, but dark tones have also been lightened by approx. 10%. From this it can be seen, that the more dark tones are lightened the more the contrast is reduced.
Figure 4g

Same as Fig. 4d, but sharpened again.
Figure 4h

Same as Fig. 4e, but sharpened again.
Figure 4 i

Same as Fig. 4f, but sharpened again.
Figure 4j

Same as Fig. 4g, but sharpened again.
Figure 4k

EXAMPLE 5:

Original scan
Figure 5a

Light and shade have been set. All values higher than 96% have been set to 100%.
Figure 5b

Sharpened
Figure 5c

Gradation has been adjusted, but the very dark parts in the picture have not been lightened.
Figure 5d

Grey tones of the sky have been set to 0% and the lighting has been sharpened.
Figure 5e

Sharpened a second time
Figure 5f

EXAMPLE 6:

Original scan
Figure 6a

Light and shade have been set.
Figure 6b

Sharpened
Figure 6c

Gradation adjusted
Figure 6d

Sharpened a second time
Figure 6e

EXAMPLE 7:

Original scan
Figure 7a

Light and shade have been set.
Figure 7b

Sharpened
Figure 7c

Standard gradation
Figure 7d

Same as 7d, but the medium tones have also been lightened by approx. 15% and the dark tones by approx. 5%.
Figure 7e

Same as 7e, but sharpened again.
Figure 7f

EXAMPLE 8:

Original scan
Figure 8a

Light and shade have been set. In contrast to Fig. 7b all tones greater than 95% have been set to 100%.
Figure 8b

Sharpened
Figure 8c

Gradation has been adjusted. Dark tones have not been lightened. This picture shows, that with deep black a high-contrast image can be obtained.
Figure 8d

Sharpened a second time
Figure 8e

EXAMPLE 9:

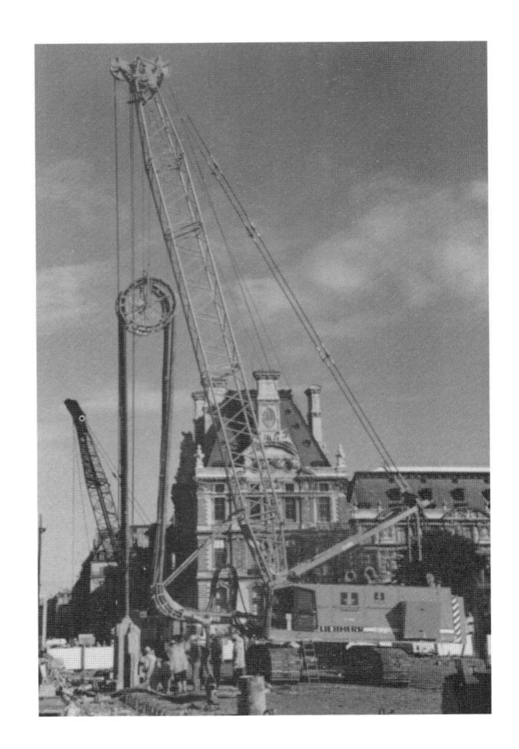

Top: original scan
Bottom left: Light and shade have been set.
Bottom right: sharpened

Top: standard gradation
Bottom left: sharpened a second time
Bottom right: Same as on the left, but the gradation has been slightly changed.

EXAMPLE 10:

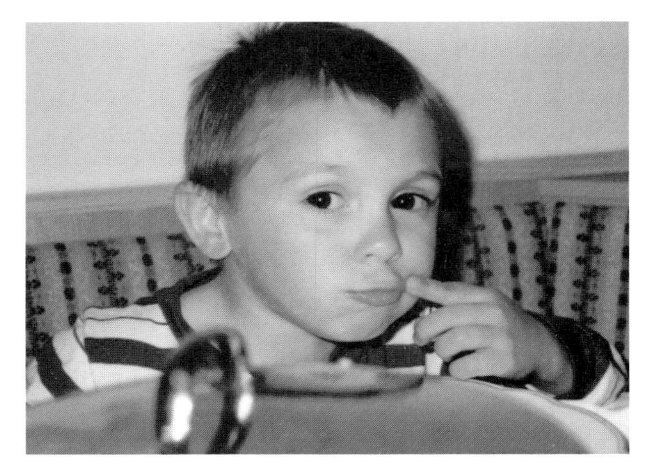

Original scan
Figure 10a

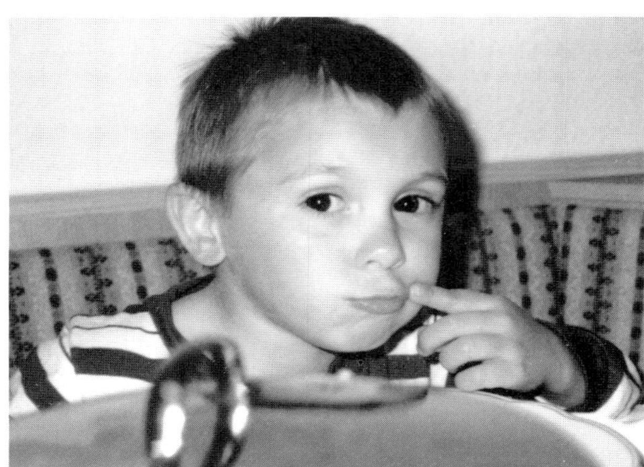

Light and shade have been set.
Figure 10b

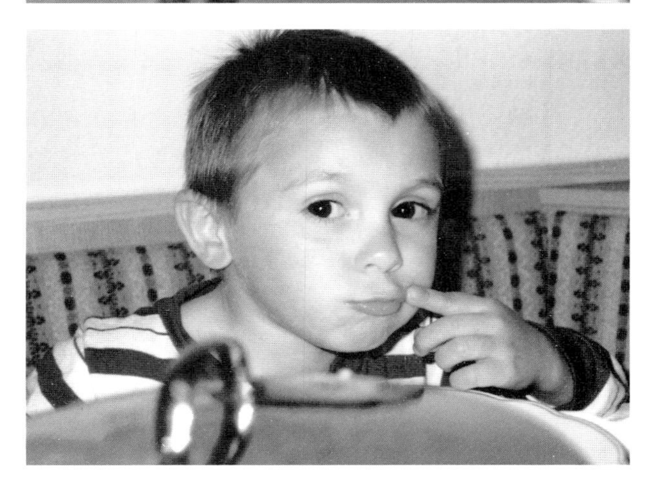

Sharpened
Figure 10c

Standard gradation
Figure 10d

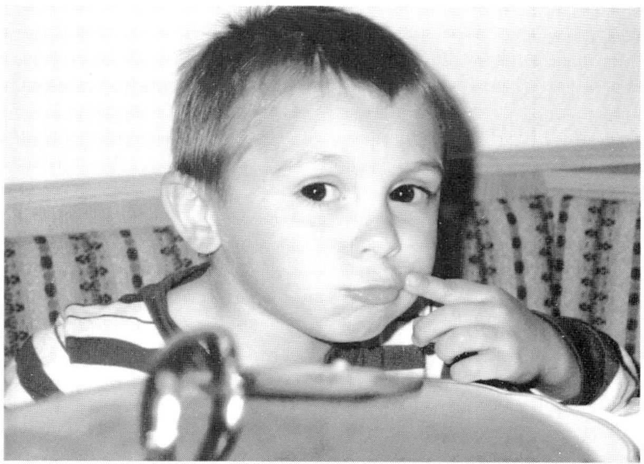

Gradation manually adjusted
Figure 10e

Sharpened a second time
Figure 10f

EXAMPLE 11:

Original scan
Figure 11a

Light and shade have been set.
Figure 11b

Sharpened
In addition, the white patch on the roof has been eliminated.
Figure 11c

Standard gradation
Figure 11d

Gradation manually adjusted
Figure 11e

Sharpened a second time
Figure 11f

EXAMPLE 12:

Original scan
Figure 12a

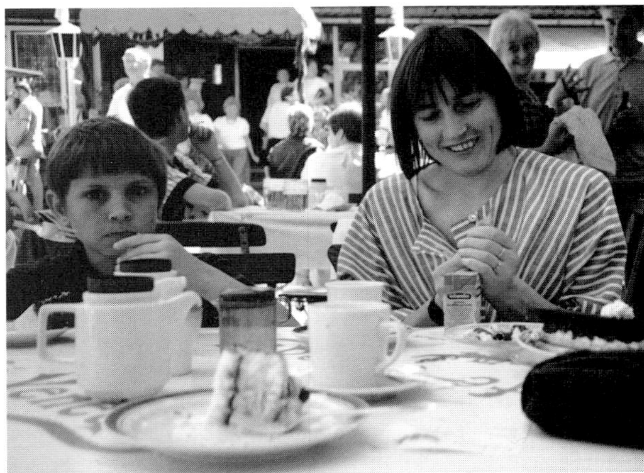

Light and shade have been set, sharpened, standard gradation
Figure 12b

Gradation manually adjusted, sharpened again
Figure 12c

EXAMPLE 13:

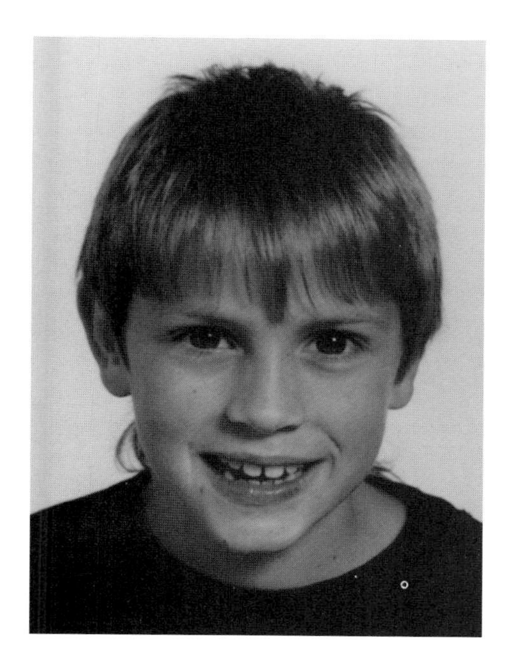

Top: original scan
Bottom left: light and shade have been set, sharpened, standard gradation
Bottom right: same as left, but less lightened

Miscellaneous

It is possible using image processing programs not only to sharpen pictures and change the brightness and contrast, but these programs also possess very convenient facilities for retouching pictures. However, we only wish to scratch the surface of this subject, as the facilities depend to a great extent on individual software packages.

Up till now we have simply talked about correcting errors and optimising pictures with respect to gradation (change in tone). However, gradation may also be used to obtain special effects, as any grey level may be assigned another different grey level. For example, a picture may be inverted.

Using the retouching function, manipulations may be carried out on the picture which conventional technology is not able to achieve. Parts of the picture may be cut out, copied, rotated and distorted.

Using gradation a picture may also be adjusted so that it is inverted.

The transition between the changed parts of the picture may be matched to the main picture, again using a smoothing filter. Pictures may be freed and then put on to another background. The most diverse effects may be obtained by using filtering, changes in tone and reduction of grey levels. In addition, text and many other enhancements may be added.

Even development demands are made on the manufacturers

As you can see, through the technological state of the art and by means of very reasonable processing aids pictures of outstanding quality may be obtained. In spite of this there is still much to be desired.

Above all, the operation of programs must be simplified considerably. We do not believe that the current philosophy of scanner and image processing programs is at its optimum. Why must DTP users conduct a running battle with scanning and image resolutions, interpolation, tone increase, moiré effects, etc.? The information contained within this book could be included within the software.

Just the gradation was changed.

There are virtually no limits to manipulation.

A new car is quickly constructed.

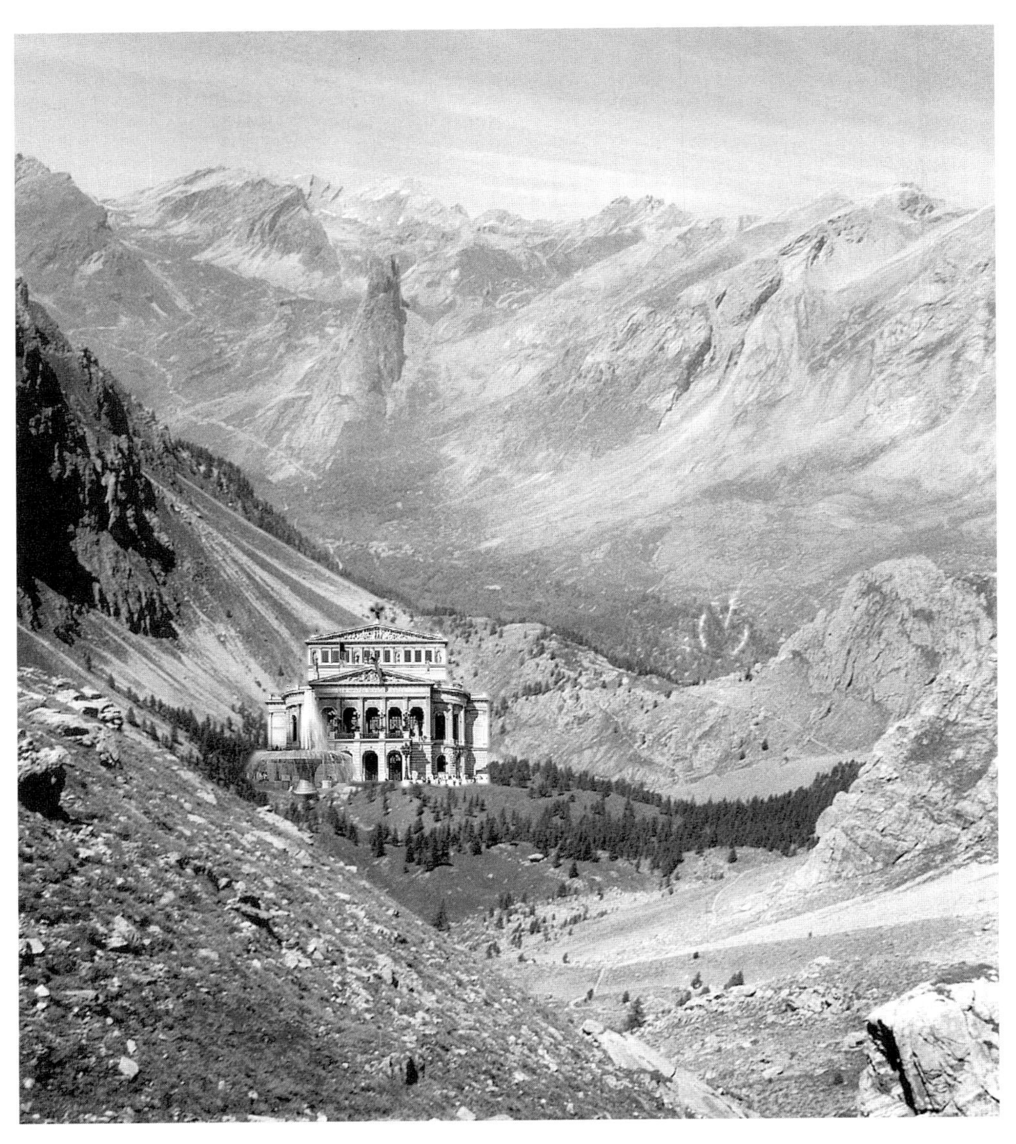

The old opera house may be found in the Italian mountains. This is the proof!

Chapter 11

Pictures from video cameras

Multi-media is the keyword

Pictures from a video camera, television, film camera, magazine, drawing program – it does not matter where they come from: all can be are loaded into the computer and processed further.

Video technology is especially fascinating. It is possible to take a series of pictures, select one, load it into the computer and almost immediately it has been integrated into the DTP document. If a suitable printer is used, the picture may even be printed in colour. Gone are the days spent waiting for prints of the pictures. The scanner, therefore, now faces competition and is no longer the only input device for the computer. This video/PC coupling may have existed for some time, but in the interim the cost and capabilities of the systems available have become more attractive.

In order to read video images the PC requires a plug-in board with a video input. Through this plug-in board the video signal is digitised and then saved on to hard disk. However, from the outset one thing must be made clear. With video technology it is not possible to attain the high resolution offered by scanners.

Still video cameras

There should be no need here to describe standard video cameras in detail. However, we would like to introduce the so-called 'still video camera'. This camera operates just like a normal camera. The difference is that film no longer needs to be photoset, but the picture data is stored on to disk instead. Up to 50 pictures may be placed on one diskette, deleted and replaced. The pictures may be viewed immediately on television, or on the computer if a digitising card has been installed.

Notes on quality

If you wish to print video pictures, you should have some idea regarding the quality of the pictures. Therefore, first of all we consider a television picture.

The PAL television standard has a vertical resolution of 625 lines, of which only 574 lines are visible, however. The width of the picture is set in the ratio of about 4 to 3 with respect to the height. From the 574 lines (three parts, vertical) it may be calculated that a horizontal resolution of 764 dots (four parts) is obtained.

Therefore a visible resolution of 574 lines × 764 dots is obtained.

> Let us digress briefly into television technology in order to explain this in more detail:
>
> For a picture 625 lines × 833 dots = 500 000 picture dots need to be transmitted. As 25 pictures (50 half pictures) are transmitted per second, this results in 500 000 picture dots multiplied by 25 pictures per second equals 13 million dots which need to be transmitted each second. If successive picture dots are changed from black to white, a frequency of 13 million dots per second/2 = 6.5 MHz would need to be transmitted.
>
> In practice this extreme case is obviously rare, so it is acceptable if the pictures are slightly unsharp occasionally. Therefore a maximum frequency of 4.125 MHz was set for the brightness (luminance). You must bear in mind here that with a video signal it is an analogue signal and not a digital signal which is being processed. If the brightness does not change significantly, then the frequency for the brightness is also relatively low.

The vertical resolution (image height) of 574 visible lines is fixed by the technology. However, considerable differences in quality occur in the horizontal resolution (width of the picture). The 764 dots specified by the television standard are not attained in practice. This means that sharp edges (abrupt changes from white to black) are shown a little blurred.

The horizontal resolution in televisions and videos is not, however, specified in dots, but instead in double lines. A white and a black line result in a double line, so that out of the 764 dots 382 double lines are produced. However, for the sake of simplicity these are simply known as lines.

In general, the horizontal resolution is specified as a frequency in megahertz. Hence a television has a horizontal resolution of about 4.4 MHz which may decrease to as low as about 3.5 MHz. Here 4.4 MHz corresponds to about 350 (double) lines and 700 dots. The calculation for the resolution is quite simple: Resolution in lines = resolution in megahertz × 80. Simple video devices often only have a horizontal resolution of about 2 MHz. This means that fine details in a picture with considerable variations in brightness are not shown so sharply.

The following horizontal resolutions are specified for the video range:

VHS:	3.2 MHz (250 lines, 500 dots)
S-VHS:	5.0 MHz (400 lines, 800 dots)
Video 8:	3.4 MHz (270 lines, 540 dots)
Hi8:	5.4 MHz (430 lines, 860 dots)

These values represent the upper limits for systems, which at present are not attained by video devices. Standard camcorders are currently still a long way off from attaining these values. The vertical resolution (image height), however, is in every case always 574 lines.

The following horizontal resolutions are attained at present:

S-VHS system:	3.9 MHz (310 lines, 620 dots)
Hi8 system:	3.9 MHz (310 lines, 620 dots)
Video 8/VHS:	2.0 MHz (160 lines, 320 dots)

(State: September 1991)

Calculation of raster width and image size

We have already mentioned which resolution a video signal can attain. As the video signal is an analogue signal it is digitised by a control board (video digitiser) installed in the PC.

For the printing of video images, it is the resolution attainable by the digitising card (video digitiser) which is of primary importance and not the resolution of the video signal. Professional video digitisers may attain, for example, a resolution of 680 × 512 dots at a colour resolution of 16.7 million colours. Independent of the quality of the video signal scanning currently always takes place at the same resolution, so that the same number of colours or grey levels always exists. Accordingly, the same technical resolution is also always used.

The quality of the picture, however, depends on the resolution of the video signal present. If the video signal received is of poor quality, then the pictures will be blurred and depth will be missing from them. However, as we will come to appreciate, this may also be improved.

Maximum image size

On the basis of the resolution of the video digitiser the maximum image size, which may be attained without special image processing, will now be calculated. In the following example a printing quality based on a raster width of 110 lpi (44 L/cm) will be assumed. As mentioned already, the video digitiser in this example has a resolution of 680 dots (width) × 512 dots (height). Hence the image size is calculated as follows:

$$Max.\ image\ size\ (cm) = \frac{resolution\ of\ video\ digitiser\ (dpi)\ \times 2.54}{raster\ width\ (lpi)\ \times 1.2}$$

Example:

$$Max.\ Bild\ width = \frac{680\ dots \times 2.54}{110\ lpi \times 1.2} = 13.0\ cm$$

$$Max.\ image\ height = \frac{512\ dots \times 2.54}{110\ lpi \times 1.2} = 9.8\ cm$$

Therefore, at a raster width of 110 lpi the video image may be printed at a size of 13 cm × 10 cm.

Maximum raster width

Using the following formula the maximum raster width, attainable without special image processing, may be calculated on the basis of the desired image size. In this example a picture with a width of 9 cm is to be printed.

$$Max.\ raster\ width\ (lpi) = \frac{resolution\ of\ video\ digitiser\ (dpi)\ \times 2.54}{image\ size\ (cm)\ \times 1.2}$$

Example:

$$Max.\ raster\ width = \frac{680\ dots \times 2.54}{9\ cm \times 1.2} = 160\ lpi\ \ (63\ L/cm)$$

Therefore, according to these calculations a picture of size 9 cm × 6.7 cm may be printed at 160 lpi (63 L/cm). However, for technical reasons the raster width should never exceed 150 lpi (59 L/cm) at a photosetter resolution of 2540 dpi.

delivered from the suppliers, are sent using the computer via satellite for example, from Spain to Germany, and then printed.

The process here is very simple. The video camera is connected to the computer and the pictures are viewed on the monitor. Pictures which are to be transmitted are 'frozen', i.e. the desired pictures are held on the monitor. The momentary image created this way may now, if necessary, be changed. The picture is then compressed and stored in a database. Once all the pictures have been selected, the desired receiving partner is selected from an address file and the data connection is automatically opened. Any transmission facility such as a telephone (modem) or data line (Datex-P) may be used.

At the receiving location the computer automatically receives the pictures and stores them on to hard disk. Therefore, the picture taken in Spain can be viewed a few minutes later in Germany. As the pictures are compressed for transmission, when they are called up decompression take place automatically. Using a video printer or a sublimation printer, paper output of outstanding quality may be reproduced immediately.

Many types of applications exist for such communication devices. Consider, for example, a journalist who wishes to send text and pictures over any distance to his or her editorial department via the telephone line.

Video standards

When working with video technology various standards need to be taken into account. This ensures that devices are compatible. Pictures which were taken according to a particular standard must also be played back by devices using the same standard. In general, professional video digitisers may process all standards.

PAL

The PAL television standard was developed in Germany and has 625 lines with 50 half pictures per second. This television standard is used worldwide in, for example, Australia, Austria, Brazil, Finland, Great Britain, India, Indonesia, Ireland, Israel, Italy, Kenya, Malta, New Zealand, Nigeria, Norway, Singapore, South Africa, Spain, Sri Lanka, Sweden, Switzerland, Tanzania, Turkey and Yugoslavia.

NTSC

The NTSC standard is the oldest colour television standard in the world and has been used in the USA since 1953. This standard has 525 lines and 60 half pictures per second. It is used nowadays in North America, South America, Japan and other Asian countries.

SECAM

The SECAM standard was developed in France and has 625 lines with 50 pictures per second. This standard is used in France and mostly in East European countries.

VHS, Video 8, S-VHS, Hi8

The VHS and Video 8 standards are particular methods for recording on to video tapes and must not be confused with the television standards. Each of these devices provides a suitable video signal (e.g. PAL FBAS signal) which may then be processed further.

FBAS

The FBAS signal is a complete colour video signal which is in general simply known as the video signal. All FBAS signals for the same television standard are identical and may be reproduced on any television screen, regardless of whether the signal has come from a VHS or a Video 8 recorder.

Y/C or Hosiden signal

The Y/C or Hosiden signal is a video signal which, for reasons of quality, feeds the individual image components (luminance, chrominance, audio) via separate lines. With a normal video signal (FBAS signal) all signals are fed via a single line.

The chrominance signal contains information on colour and the luminance signal contains information on the brightness. The audio signal is the sound.

HDTV (High Definition Television)

The HDTV standard is currently being developed in Japan and Europe and will become the new television standard. This standard has a much better image representation with a screen ratio of 3:5 and a vertical image resolution of over 1000 lines. However, inexpensive televisions and video recorders will probably not be on the market until the end of the 1990s.

Chapter 12
Scanning and printing of pictures: summary and tables

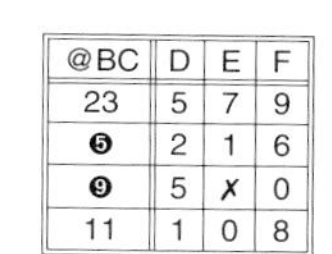

This chapter should provide an overview of the subject of scanning and printing. In addition, the most important tables are given here.

General information on the scanning of greyscale pictures

As already described, with respect to the reproduction of grey tones, half-toning is performed by devices capable only of printing black dots. If a printed picture is examined closely using a magnifying glass it will be seen that the picture comprises many individual dots. Each grey dot in the picture is transformed into several or larger dots. Light grey tones are represented by small dots and dark grey tones by larger dots.

However, as many output devices are only capable of printing dots (pixels) of a fixed size, dots of varying sizes are attained by combining several pixels together.

Therefore, if you wish to print a picture with 100 grey levels, 10×10 pixels are required in order to print a single raster dot with 100 different sizes. To represent 100 grey levels, the printer must have a resolution which is about ten times higher than that of the scanned picture and for 25 grey levels it must have a resolution which is about five times higher. It is therefore completely worthless scanning at a high resolution, if subsequent printing is simply carried out by a laser printer. This merely results in a large computer memory requirement and an unnecessarily long printing time.

The printing quality of photographs (half-tone pictures) is determined both by the number of raster dots and the number of grey levels. The number of raster dots is generally known as 'raster width', 'half-tone raster', 'raster frequency' or 'grid raster'. It is measured in lines per inch (lpi) or lines per cm (L/cm). The selected raster width

@BC	D	E	F
23	5	7	9
❺	2	1	6
❾	5	✗	0
11	1	0	8

depends on the resolution of the output device (photosetter), paper quality and printing process (printing medium).

In general, however, raster widths of 140 lpi (55 L/cm) or 150 lpi (60 L/cm) are used for high-quality prints. Larger raster widths are normally not recommended and in any case are only worth using with the new photosetters which have resolutions of more than 3300 dpi (1300 L/cm).

The image resolution and hence the scanning resolution should be calculated according to the formulae given in this book. If the calculated scanning resolution is less than half the physical resolution of the scanner, then scanning should be carried out at a higher resolution with subsequent data reduction. It is recommended always to scan at the optical resolution of the scanner and then to carry out a suitable data reduction. However, with data reduction a facility for interpolating the grey levels must exist.

Pictures must not be rasterised (or dithered) by the scanner. Rasterising is carried out later by means of the printer, photosetter (RIP) or DTP program. If pictures have already been rasterised (dithered) by the scanner, then changing the size of the picture at a later stage results in a loss in quality (e.g. moiré effects). The values and notes given in this book are only valid for non-rasterised scanning! In addition, pictures must be stored in the TIFF format before further processing.

If rasterised original documents (pictures from magazines, brochures, etc.) are subsequently scanned, moiré effects are produced which should be removed or reduced by using special measures. In this case scanning must be carried out at the physical resolution. The picture then needs to be smoothed and the resolution reduced as required. When the data is reduced, the grey levels should be interpolated.

When using offset printing pictures become much darker due to the inherent tone increase. This tone increase should be taken into account either when scanning or when image processing. Pictures may also become darker during photosetting.

In order to adjust the pictures on the screen, the monitor must be capable of representing grey levels and should have been calibrated. The resolution of the monitor for greyscale pictures should be at least 640 × 400 dots.

Drawing and retouching

@BC	D	E	F
23	5	7	9
❺	2	1	6
❾	5	✗	0
11	1	0	8

Three types of picture data or image processing programs exist:

- pixel pictures
- vector pictures
- greyscale pictures

If a line drawing is scanned, a pixel picture is obtained first. This picture may be processed further using suitable pixel-orientated programs, such as Paintbrush or IMAGE Plus. Often simple image processing functions are incorporated within the scanner software.

If a raster/vector conversion (auto-tracing) is performed on the scanned line drawing, the converted drawing may then be processed further using a vector-orientated program. Artline or Corel Draw are drawing programs which are capable of carrying out a semi-automatic conversion, as well as being convenient facilities for further processing.

If a greyscale picture is scanned the data must not be rasterised. At this point all the greyscale information is available. Therefore, it is not possible to process the picture using a pixel-orientated drawing program. A greyscale picture must be processed using a suitable image processing program which can also represent real grey levels. As long as scanner software is incapable of representing real grey levels, and hence not suitable for further processing, an image processing program is definitely recommended for the optimum adjustment of pictures. Picture Publisher or Image Studio are examples of programs which are capable of representing greyscale pictures and processing TIFF files further.

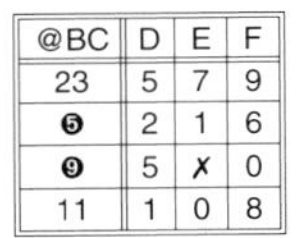

Checklist for greyscale scanning

❒ Type of scanner

The quality attainable for greyscale pictures depends mainly on how many grey levels the scanner is able to differentiate. Good-quality pictures may be attained with a 64 greyscale scanner, but a scanner with 264 grey levels is necessary for professional-quality pictures. The optimum scanners are those which operate internally with 1024 or 4096 grey levels.

❒ Basic scanner setting

If a greyscale picture is scanned it must not be rasterised by the scanner. Accordingly, real greyscale information is saved. The various settings which need to be made depend on the type of scanner. Either 'rasterising' or 'dithering' needs to be set to 'off' or 'grey levels' or 'greyscale' should be selected. In addition, an average setting (50% or 0%) should be used for brightness and contrast. Picture manipulation is carried out mainly through the gradation setting.

❒ Data format

The TIFF format should be used as the data format as it is the only format which saves real grey levels. Under no circumstances should you save or convert using, for example, the PCX or IMG format.

If the program used to process the pictures further is not capable of reading the TIFF format, then it is possible to convert the data into the PostScript format (EPS). If your scanner does not provide this facility, you must use a suitable conversion program.

❒ Scanning and image resolution

The scanning resolution should be set according to the formulae or tables given in this book. In principle, the scanning resolution may be set higher than calculated, but the memory requirement is then increased considerably. A doubling of the scanning resolution results in four times the amount of memory being required! However, the quality of pictures may be improved if scanning is carried out at a higher (max. optical) resolution and then, using an image processing program, data reduction is carried out later by means of interpolating the grey levels.

❒ Quality of the original documents

If the original document is not a photograph, but a picture which has already been rasterised, then moiré effects are produced when scanning. With rasterised originals scanning should be carried out at the physical resolution of the scanner. Further processing of the picture (smoothing, data reduction, interpolation) should then be carried out.

@BC	D	E	F
23	5	7	9
❺	2	1	6
❾	5	✗	0
11	1	0	8

- ❐ Image processing

  Essentially, greyscale pictures should only be evaluated on a greyscale screen and not through a printout from a laser printer. The most important consideration in image processing is that the tone increase, which is produced on printing and perhaps also on photosetting, is taken into account. To enable pictures to be adjusted on the monitor it should be calibrated beforehand.

  In order to obtain pictures which do not lack contrast, the lightest grey level should be set to 5% and the darkest to 95%. Depending on the picture composition the greyscale range may also be set from 0% to 100%.

- ❐ Picture settings within the DTP program

  Both the raster width and the raster angle must be selected in the DTP program. The angle of the raster is always 45° for black and white pictures. The raster width depends on the quality of paper used and is 150 lpi (60 L/cm) for high-quality pictures.

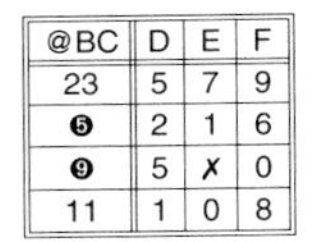

@BC	D	E	F
23	5	7	9
❺	2	1	6
❾	5	✗	0
11	1	0	8

Quality of paper

The maximum raster width which may be used is dependent on the output device, the printing medium and the quality of paper. As described already, the tone increase also depends on the quality of paper. The lower the quality of paper the higher the tone increase.

With respect to the tone increase, the quality of paper may be divided into three classes:

- *Class 1, art printing paper*: All shiny and semi-matt coated paper with a weight higher than 70 g/m^2. Coated matt paper over 70 g/m^2 is classified as class 1 or class 2 depending on the quality of the printed result.
- *Class 2, picture printing paper*: All coated paper with a weight less than 70 g/m^2.
- *Class 3, natural paper*: All uncoated paper (including pigmented/satin coated paper).

The quality of paper may be classified further depending on the raster widths which may be used:

- Coated paper for raster widths of over 140 lpi (55 L/cm)
- Satin coated paper for raster widths between 100 lpi (40 L/cm) and 140 lpi
- Uncoated, machine-smooth paper for raster widths between 75 lpi (30 L/cm) and 100 lpi

The following raster widths are normally used:

- 150 lpi (60 L/cm) or 140 lpi (55 L/cm) for high-quality reproduction on to picture printing paper (coated paper)
- 120 lpi (47 L/cm) for average-quality reproduction
- 75 lpi (29 L/cm) for newspapers

Recommended raster widths, 45° angle

@BC	D	E	F
23	5	7	9
❺	2	1	6
❾	5	✗	0
11	1	0	8

The following tables show the standard raster widths, the printer or photosetting resolution required for them and the number of grey levels attainable. The image size has not yet been taken into account for the given image resolutions. Either the image size is entered separately by means of the software or the resolution must be matched to correspond with the size.

Recommended raster width for photosetter and laser printer

Photosetter/printer resolution	Raster width 45°	Grey levels	Image resolution
300 dpi	53 lpi (21 L/cm)	32	74 dpi
400 dpi	70 lpi (28 L/cm)	32	99 dpi
600 dpi	106 lpi (42 L/cm)	32	148 dpi
"	85 lpi (33 L/cm)	50	119 dpi
"	71 lpi (28 L/cm)	72	90 dpi
635 dpi	112 lpi (44 L/cm)	32	157 dpi
"	90 lpi (35 L/cm)	50	126 dpi
"	75 lpi (29 L/cm)	72	105 dpi
"	64 lpi (25 L/cm)	98	90 dpi
813 dpi	96 lpi (38 L/cm)	72	134 dpi
"	82 lpi (32 L/cm)	98	115 dpi
"	72 lpi (28 L/cm)	128	101 dpi
846 dpi	100 lpi (39 L/cm)	72	140 dpi
"	85 lpi (34 L/cm)	98	120 dpi
"	75 lpi (29 L/cm)	128	105 dpi
900 dpi	106 lpi (42 L/cm)	72	148 dpi
"	91 lpi (36 L/cm)	98	127 dpi
"	80 lpi (31 L/cm)	128	111 dpi
"	71 lpi (28 L/cm)	162	99 dpi
1000 dpi	118 lpi (46 L/cm)	72	164 dpi
"	101 lpi (40 L/cm)	98	141 dpi
"	88 lpi (35 L/cm)	128	124 dpi

@BC	D	E	F
23	5	7	9
❻	2	1	6
❾	5	✗	0
11	1	0	8

Photosetter resolution	Raster width 45°	Grey levels	Image resolution
1000 dpi	79 lpi (31 L/cm)	162	110 dpi
"	71 lpi (28 L/cm)	200	99 dpi
1200 dpi	121 lpi (48 L/cm)	98	170 dpi
"	106 lpi (42 L/cm)	128	148 dpi
"	94 lpi (37 L/cm)	162	132 dpi
"	85 lpi (33 L/cm)	200	119 dpi
"	77 lpi (30 L/cm)	242	108 dpi
1250 dpi	126 lpi (50 L/cm)	98	177 dpi
"	110 lpi (43 L/cm)	128	155 dpi
"	98 lpi (39 L/cm)	162	137 dpi
"	88 lpi (34 L/cm)	200	124 dpi
"	80 lpi (32 L/cm)	242	112 dpi
1270 dpi	128 lpi (51 L/cm)	98	179 dpi
"	112 lpi (44 L/cm)	128	157 dpi
"	100 lpi (39 L/cm)	162	140 dpi
"	90 lpi (35 L/cm)	200	126 dpi
"	82 lpi (32 L/cm)	242	114 dpi
1448 dpi	146 lpi (58 L/cm)	98	205 dpi
"	128 lpi (50 L/cm)	128	179 dpi
"	114 lpi (45 L/cm)	162	159 dpi
"	102 lpi (40 L/cm)	200	143 dpi
"	93 lpi (37 L/cm)	242	130 dpi
"	85 lpi (34 L/cm)	256	119 dpi
1524 dpi	156 lpi (61 L/cm)	98	218 dpi
"	136 lpi (54 L/cm)	128	191 dpi
"	121 lpi (48 L/cm)	162	170 dpi
"	109 lpi (43 L/cm)	200	153 dpi
"	99 lpi (39 L/cm)	242	139 dpi
"	91 lpi (36 L/cm)	256	127 dpi
1600 dpi	162 lpi (64 L/cm)	98	226 dpi

Photosetter resolution	Raster width 45°	Grey levels	Image resolution
1600 dpi	141 lpi (56 L/cm)	128	198 dpi
"	126 lpi (49 L/cm)	162	176 dpi
"	113 lpi (45 L/cm)	200	158 dpi
"	102 lpi (41 L/cm)	242	144 dpi
"	94 lpi (37 L/cm)	256	132 dpi
1693 dpi	150 lpi (60 L/cm)	128	209 dpi
"	133 lpi (52 L/cm)	162	186 dpi
"	120 lpi (47 L/cm)	200	168 dpi
"	109 lpi (43 L/cm)	242	152 dpi
"	100 lpi (39 L/cm)	256	140 dpi
"	92 lpi (36 L/cm)	256	129 dpi
1778 dpi	157 lpi (62 L/cm)	128	220 dpi
"	140 lpi (55 L/cm)	162	195 dpi
"	126 lpi (50 L/cm)	200	176 dpi
"	114 lpi (45 L/cm)	242	160 dpi
"	105 lpi (41 L/cm)	256	147 dpi
"	97 lpi (38 L/cm)	256	135 dpi
1800 dpi	159 lpi (63 L/cm)	128	223 dpi
"	141 dpi 56 (L/cm)	162	198 dpi
"	127 lpi (50 L/cm)	200	178 dpi
"	116 dpi 46 (L/cm)	242	162 dpi
"	106 lpi (42 L/cm)	256	148 dpi
"	98 lpi (39 L/cm)	256	137 dpi
2000 dpi	178 lpi (70 L/cm)	128	247 dpi
"	157 lpi (62 L/cm)	162	220 dpi
"	141 lpi (56 L/cm)	200	198 dpi
"	129 lpi (51 L/cm)	242	180 dpi
"	118 lpi (46 L/cm)	256	165 dpi
"	109 lpi (43 L/cm)	256	152 dpi
"	101 lpi (40 L/cm)	256	141 dpi

@BC	D	E	F
23	5	7	9
❺	2	1	6
❾	5	✗	0
11	1	0	8

@BC	D	E	F
23	5	7	9
❺	2	1	6
❾	5	✗	0
11	1	0	8

Photosetter resolution	Raster width 45°	Grey levels	Image resolution
2000 dpi	94 lpi (37 L/cm)	256	132 dpi
2032 dpi	160 lpi (63 L/cm)	162	223 dpi
"	144 lpi (63 L/cm)	200	201 dpi
"	131 lpi (51 L/cm)	242	183 dpi
"	120 lpi (47 L/cm)	256	168 dpi
"	111 lpi (44 L/cm)	256	155 dpi
"	103 lpi (40 L/cm)	256	144 dpi
"	96 lpi (38 L/cm)	256	134 dpi
2400 dpi	170 lpi (67 L/cm)	200	238 dpi
"	154 lpi (61 L/cm)	242	216 dpi
"	141 lpi (56 L/cm)	256	198 dpi
"	131 lpi (51 L/cm)	256	183 dpi
"	121 lpi (48 L/cm)	256	170 dpi
"	113 lpi (46 L/cm)	256	158 dpi
"	106 lpi (42 L/cm)	256	148 dpi
"	100 lpi (39 L/cm)	256	140 dpi
2540 dpi	163 lpi (64 L/cm)	242	228 dpi
"	150 lpi (59 L/cm)	256	209 dpi
"	138 lpi (54 L/cm)	256	193 dpi
"	128 lpi (51 L/cm)	256	179 dpi
"	120 lpi (47 L/cm)	256	168 dpi
"	112 lpi (44 L/cm)	256	157 dpi
"	106 lpi (42 L/cm)	256	148 dpi
"	100 lpi (39 L/cm)	256	140 dpi
3048 dpi	180 lpi (71 L/cm)	256	251 dpi
"	166 lpi (65 L/cm)	256	232 dpi
"	154 lpi (60 L/cm)	256	215 dpi
3556 dpi	210 lpi (82 L/cm)	256	293 dpi
"	193 lpi (76L/cm)	256	271 dpi
"	180 lpi (71 L/cm)	256	251 dpi

Calculation of actual raster widths

@BC	D	E	F
23	5	7	9
❺	2	1	6
❾	5	X	0
11	1	0	8

Laser printers and photosetters or RIPs are not capable of creating every raster width. Using the following formula the actual raster width which the current output device employs may be calculated. However, the exact value does not need to be entered into the DTP software. It is sufficient if an approximate value is used for the raster width. Standard values may be looked up in the previous tables.

$$\text{Raster width (at a raster angle of } 45^{o}) \approx \frac{\text{photosetter resolution}}{1.414\,x}$$

(This formula may only be used for a raster angle of 45°.) Any whole number may be substituted for x. For example, $x = 12$ results in a raster width of 150 lpi. For finer rasters a smaller value needs to be substituted for x. For laser printers the value 4 or larger may be used.

Example:

$$\text{Raster width (angle of } 45^{o}) \approx \frac{2540}{1.414 \times 12} \approx \frac{2540}{16.968} \approx 149.69 \text{ lpi}$$

However, instead of entering the exact value of 149.69 into the DTP software the rounded value of 150 lpi is entered.

Note:
The variable x may be explained as follows. The photosetter may only arrange whole pixels strictly in the direction of photosetting (x) and in the direction of feed (y) of the film or paper. Therefore, the variable x is given by the number of pixels arranged in the x direction. As an angle of 45° is always used in black and white reproductions the number of pixels set in the y direction is identical to the number of pixels set in the x direction. For this reason the above formula uses only x and not both x and y for calculation. The variable x represents one side of a right-angled triangle. The hypoteneuse of this triangle is thus the raster width. The general formula for any raster width is as follows:

$$\text{Raster width} = \frac{\text{photosetter resolution}}{\sqrt{(x^2 + y^2)}}$$

At a raster angle of 45° x and y are identical. This results in the following formula:

$$\text{Raster width} = \frac{\text{photosetter resolution}}{\sqrt{(x^2 + x^2)}} = \frac{\text{photosetter res.}}{\sqrt{(2 \times x^2)}} \approx \frac{\text{photosetter res.}}{1.414\,x}$$

$$\text{Image resolution} = \text{raster width} \times 1.4 \times \frac{\text{desired size}}{\text{size of original}}$$

@BC	D	E	F
23	5	7	9
❺	2	1	6
❾	5	✗	0
11	1	0	8

Formulae for greyscale pictures

In the following section you will find formulae for calculating the image resolution. The scanning resolution should be higher (max. optical resolution) for reasons of quality. Using an image processing program the data may then be reduced to the required image resolution. If the scanner carries out interpolation internally, then image resolution and scanning resolution may be identical.

$$\textit{Image resolution} = \textit{raster width} \times 1.4 \times \frac{\textit{desired size}}{\textit{size of original}}$$

This formula is valid if the image size is taken into account via the resolution, i.e. the image size is not changed. Instead of the factor 1.4 above the factor 1.2 may be sufficient.

$$\textit{Image resolution} = \textit{raster width} \times 1.4$$

This formula is valid when the image size is entered separately. If necessary, the factor 1.2 instead of 1.4 is sufficient.

$$\textit{Image size in \%} = \frac{\textit{desired size}}{\textit{size of original}} \times 100$$

$$\textit{Number of grey levels} = \left(\frac{\textit{photosetter resolution}}{\textit{raster width}}\right)^2$$

Conversion of inches into centimetres

Length in inches = length in cm/2.54

Length in cm = length in inches × 2.54

Resolution in inches = resolution in cm × 2.54

Resolution in cm = resolution in inches/2.54

The printing order

@BC	D	E	F
23	5	7	9
❺	2	1	6
❾	5	✗	0
11	1	0	8

The values tabulated below give the tone increase for various qualities of paper according to the FOGRA standard. 'FOGRA' is the German research association for printing and reproduction technology. As already mentioned, research associations in Great Britain and the United States independently attained approximately the same results. You should inform the printing firm which values it needs to comply with. The tone increase is always specified with respect to a tone on the original film of 40% and 80% (nominal values) at a raster width of 150 lpi (60 L/cm). In addition, a measurement at 50% with a raster width of 135 lpi (35 L/cm) is sometimes used. Both types of measurement process are almost identical. The tone increase which may arise during photosetting is not taken into account in this table.

Page reversed positive camera-ready copy film is normally used in Europe.

Tone increase when printing to FOGRA
Raster width: 150 lpi (60 L/cm)
Positive photosetting

The specifications are with respect to a raster width of 150 lpi (60 L/cm).
At lower raster widths the tone increase is lower.
The colour used for printing is black.

Tone of film		Tone increase which occurs in printing			
135 lpi	150 lpi	Special case*)	Paper class 1	Paper class 2	Paper class 3
50 %	**40 %**	15 %	19 %	22 %	25 %
	50 %	16 %	20 %	23 %	26 %
	70 %	13 %	17 %	20 %	23 %
	75 %	12 %	16 %	18 %	21 %
	80 %	10 %	14 %	16 %	18 %

*) Special case: modern printing machines have a lower tone increase than that recognised by FOGRA. For this reason these values were also added to the table.

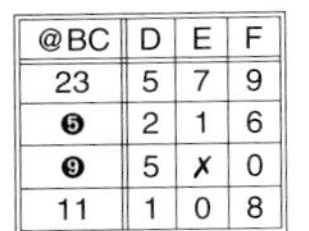

Tone increase when printing to FOGRA
Raster width: 150 lpi (60 L/cm)
Negative photosetting

The specifications are with respect to a raster width of 150 lpi (60 L/cm).
At lower raster widths the tone increase is lower.
The colour used for printing is black.

Tone nominal (film)		Tone increase which occurs in printing			
135 lpi	150 lpi	Special case*)	Paper class 1	Paper class 2	Paper class 3
50 %	**40 %** (60%)	22 %	26 %	29 %	32 %
	50 % (50%)	23 %	27 %	30 %	33 %
	70 % (30%)	19 %	23 %	26 %	29 %
	75 % (25%)	17 %	21 %	23 %	26 %
	80 % (20%)	14 %	18 %	20 %	22 %

*) Special case: modern printing machines have a lower tone increase than that recognised by FOGRA. For this reason these values were also added to the table.

The values given should usually be adhered to within a tolerance of less than ± 5% for 68% of the page. Lower tolerances must be agreed upon with the printing firm. Usually the 40% and 80% values (nominal values) are measured at a raster width of 150 lpi (60 L/cm). However, measurement may also be carried out commonly with a grey level of 50% at a raster width of 135 lpi (35 L/cm).

Please note that the tone of 40% will become 60% on the film when the film is photoset negative and the tone of 80% will become 20%.

Tip:
Plan to have small rasterised areas with 40%, 80% and 100% grey levels at a raster width of 150 lpi in your document as stylistic elements. At any time you can then control the tone increase which occurs during photosetting and printing.

Correction of tone increase - General

@BC	D	E	F
23	5	7	9
❺	2	1	6
❾	5	✗	0
11	1	0	8

How the tone increase above may be taken into account is described in this section. The values specified in the tables may be entered in a suitable image processing program as inputs and outputs for the gradation. The current grey level for the picture is entered as the input value and the output value is taken as the new replacement input value. The intermediate unspecified values are automatically interpolated.

Warning: In the following tables the current correction factors which may be used in the image processing program are shown and not the tone increase. In image processing the values for tone increase must not be used, only the corrections for the tone increase.

The tables may be used as follows. The printing firm, for example, specifies a tone increase of 19% (at a 40% grey level and at a raster width of 150 lpi (60 L/cm) for the selected type of paper. Accordingly, the column with a tone increase of 19% is located. The values given in this column may then be entered into the image processing program as output values. In general, it is better to take into account a tone increase which is too high than one which is too low.

In general, scanners have problems in recognising 100% values as black. Furthermore, the picture looks better if it contains very light areas as well as areas of deep black. Therefore, in contrast to the values given in the following tables, an output of 0% may be entered for an input of 0% and an output of 100% for an input of 100%.

For example, with respect to the following table the values given may be entered to the gradation for a tone increase of 19%:

Input	Output
0%	3% (or 0%)
30%	18%
60%	41%
80%	62%
90%	81%
100%	98% (or100%)

Correction of tone increase: positive photosetting

The following tables relate to positive photosetting. This means that the areas to be printed are black. A higher quality is attainable using positive rather than negative film. Hence, negative film should only be used if the printing firm does not work with positive printing plates.

For the following values it is assumed that the tone increase which occurs due to photosetting is negligible, i.e. that photosetting is carried out on to film with lith characteristic and that the photosetter is suitably calibrated.

The following table shows tone corrections for pictures with a raster width of 150 lpi. The tables which follow this relate to pictures with raster widths of 120 lpi and 75 lpi. Irrespective of the raster width at which the pictures where photoset, the 40% and 80% measurement areas should always have a raster width of 150 lpi (60 L/cm).

The special case (15%) given in these tables is not included in the standardisation, but is often required in practice.

Correction: positive / 150 lpi (60 L/cm)

Tone increase corrections, positive film, photosetter calibrated, printing colour: black

Input	Outputs for a tone increase, occurring during printing, of			
	15%*) (Special case)	19%*) (Paper class 1)	22% *) (Paper class 2)	25% *) Paper class 3
0%	3%	3%	3%	3%
30%	19%	18%	17%	15%
60%	44%	41%	39%	36%
80%	66%	62%	58%	54%
90%	82%	81%	76%	71%
100 %	98%	98%	95%	94%

@BC	D	E	F
23	5	7	9
❺	2	1	6
❾	5	✗	0
11	1	0	8

Correction: positive / 120 lpi (47 L/cm)

Tone increase corrections, positive film, photosetter calibrated, printing colour: black

Input	Outputs for a tone increase, occurring during printing, of			
	15%*) (Special case)	19%*) (Paper class 1)	22% *) (Paper class 2)	25% *) (Paper class 3)
0%	3%	3%	3%	3%
30%	20%	19%	18%	17%
60%	47%	44%	42%	40%
80%	70%	65%	62%	60%
90%	85%	84%	80%	78%
100 %	98%	98%	96%	95%

Correction: positive / 75 lpi (30 L/cm)

Tone increase corrections, positive film, photosetter calibrated, printing colour: black

Input	Outputs for a tone increase, occurring during printing, of			
	15%*) (Special case)	19%*) (Paper class 1)	22% *) (Paper class 2)	25% *) (Paper class 3)
0%	3%	3%	3%	3%
30%	24%	22%	21%	20%
60%	52%	49%	48%	47%
80%	74%	71%	70%	68%
90%	88%	86%	85%	84%
100 %	98%	98%	97%	97%

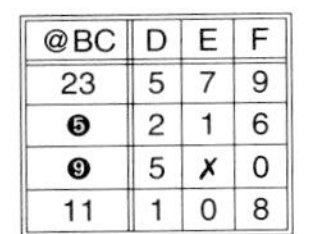

Correction of tone increase: negative photosetting

The following tables relate to negative photosetting. This means that the areas to be printed are white and those not to be printed are black. The quality attainable is lower using negative rather than positive film. Hence negative film should only be used when the printing firm does not work with positive printing plates.

For the following values it is assumed that the tone increase due to photosetting is negligible, i.e. that photosetting is carried out on to film with lith characteristic and that the photosetter is suitably calibrated.

The following table shows tone corrections for pictures with a raster width of 150 lpi. The tables which follow this relate to pictures with raster widths of 120 lpi and 75 lpi. Irrespective of the raster width at which the pictures where photoset, the 40% and 80% measurement areas should always have a raster width of 150 lpi (60 L/cm).

The special case (22%) given in these tables is not included in the standardisation, but is often required in practice.

Correction: negative / 150 lpi (60 L/cm)

Tone increase corrections, negative film, photosetter calibrated, printing colour: black

Input	Outputs for a tone increase, occurring during printing, of			
	22%*) (Special case)	26%*) (Paper class 1)	29% *) (Paper class 2)	32% *) Paper class 3
0%	3%	3%	3%	3%
30%	17%	14%	13%	12%
60%	38%	35%	33%	30%
80%	59%	54%	51%	47%
90%	77%	71%	67%	63%
100 %	98%	98%	95%	92%

@BC	D	E	F
23	5	7	9
❺	2	1	6
❾	5	✗	0
11	1	0	8

Correction: negative / 120 lpi (47 L/cm)

Tone increase corrections, negative film, photosetter calibrated, printing colour: black

Input	Outputs for a tone increase, occurring during printing, of			
	22%*) (Special case)	26%*) (Paper class 1)	29% *) (Paper class 2)	32% *) Paper class 3
0%	3%	3%	3%	3%
30%	19%	18%	17%	16%
60%	42%	40%	38%	36%
80%	64%	61%	57%	55%
90%	81%	78%	74%	72%
100 %	98%	98%	96%	96%

Correction: negative / 75 lpi (30 L/cm)

Tone increase corrections, negative film, photosetter calibrated, printing colour: black

Input	Outputs for a tone increase, occurring during printing, of			
	22%*) (Special case)	26%*) (Paper class 1)	29% *) (Paper class 2)	32% *) Paper class 3
0%	3%	3%	3%	3%
30%	21%	19%	18%	17%
60%	48%	47%	44%	42%
80%	71%	68%	67%	65%
90%	86%	84%	83%	80%
100 %	98%	98%	98%	98%

@BC	D	E	F
23	5	7	9
❺	2	1	6
❾	5	✗	0
11	1	0	8

Photosetting order

We herewith place an order with you to photoset the files listed under the specified conditions.

File names: ____

____

❑ **MS-DOS** ❑ **Macintosh** PostScript printing files

Includes greyscale pictures: ❑ No ❑ Yes Film with
Tone change <–3%/+4%
Full-tone density > log 3.0
Background veil < log 0,06)

Resolution: ❑ 635 dpi ❑ 1270 dpi ❑ 2540 dpi ❑ __________

Output: ❑ Film: page reversed/positive ❑ __________

❑ Photographic paper

Size: ❑ Half ❑ Double ❑ Letter

❑ Legal ❑ Half ❑ Broad Sheet ❑ __________

Number of pages: ________ pages

Price per page to: ❑ Size of file (KByte) ❑ Fixed price

Further conditions: ____

____

Contact: Telephone: ______________________________

Name: ______________________________

Return: ❑ Normal ❑ Courier ❑ Collect

Date/signed

Grey levels as logarithms and percentages

@BC	D	E	F
23	5	7	9
❺	2	1	6
❾	5	✗	0
11	1	0	8

Using the following tables grey levels (raster tone) which have been given as densities (logarithmic measurements) may be converted into percentages. Hence you have a conversion table readily available which you may use to convert the densities, which the printing firm uses, to percentages which you use on the computer. The first column in the table shows the density. In order to determine the exact raster tone as a percentage, the column with the relevant density for the full-tone area is selected.

Usually a full-tone density of about log 1.8 is attained for art printing paper, about log 1.4 for picture printing paper and about log 1.0 for natural paper. For film the ∞ column is used, since with films a full-tone density of about log 3.0 is attained.

Example: Using a densitometer the density of a grey area is measured as log 0.08 and the density of the full-tone area (black) as log 1.8. Locate the density of the grey area of 0.08 in the first column and then read the value 17.1% from the column with the full-tone density of 1.8.

Conversion table for density to raster tone

Density	Raster tone (grey level) as a percentage at a full-tone density of							
log	∞	2,0	1,8	1,6	1,4	1,2	1,0	0,8
0,01	2,3%	2,3%	2,3%	2,3%	2,4%	2,4%	2,5%	2,7%
0,02	4,5%	4,5%	4,6%	4,6%	4,7%	4,8%	5,0%	5,3%
0,03	6,7%	6,7%	6,8%	6,8%	7,0%	7,1%	7,4%	7,9%
0,04	8,8%	8,9%	8,9%	9,0%	9,2%	9,4%	9,8%	10,5%
0,05	10,9%	11,0%	11,1%	11,2%	11,3%	11,6%	12,1%	12,9%
0,06	12,9%	13,0%	13,1%	13,2%	13,4%	13,8%	14,3%	15,3%
0,07	14,9%	15,0%	15,1%	15,3%	15,5%	15,9%	16,5%	17,7%
0,08	16,8%	17,0%	17,1%	17,3%	17,5%	18,0%	18,7%	20,0%
0,09	18,7%	18,9%	19,0%	19,2%	19,5%	20,0%	20,8%	22,2%

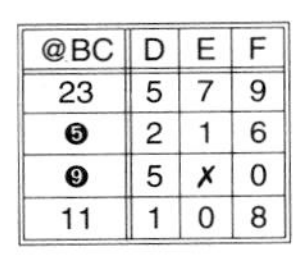

@BC	D	E	F
23	5	7	9
❺	2	1	6
❾	5	✗	0
11	1	0	8

Density	Raster tone (grey level) as a percentage at a full-tone density of							
log	∞	2,0	1,8	1,6	1,4	1,2	1,0	0,8
0,10	20,6%	20,8%	20,9%	21,1%	21,4%	22,0%	22,9%	24,4%
0,12	24,1%	24,4%	24,5%	24,8%	25,1%	25,8%	26,8%	28,7%
0,15	29,2%	29,5%	29,7%	30,0%	30,4%	31,2%	32,5%	34,7%
0,18	33,9%	34,3%	34,5%	34,8%	35,3%	36,2%	37,7%	40,3%
0,21	38,3%	38,7%	39,0%	39,3%	39,9%	40,9%	42,6%	45,6%
0,24	42,5%	42,9%	43,1%	43,5%	44,2%	45,3%	47,2%	50,5%
0,27	46,3%	46,8%	47,0%	47,5%	48,2%	49,4%	51,4%	55,0%
0,30	49,9%	50,4%	50,7%	51,2%	51,9%	53,2%	55,4%	59,3%
0,35	55,3%	55,9%	56,2%	56,8%	57,6%	59,1%	61,5%	65,8%
0,40	60,2%	60,8%	61,2%	61,7%	62,7%	64,2%	66,9%	71,5%
0,45	64,5%	65,2%	65,6%	66,2%	67,2%	68,9%	71,7%	76,7%
0,50	68,4%	69,1%	69,5%	70,1%	71,2%	73,0%	76,0%	81,3%
0,55	71,8%	72,5%	73,0%	73,7%	74,8%	76,7%	79,8%	85,3%
0,60	74,9%	75,6%	76,1%	76,8%	78,0%	79,9%	83,2%	89,0%
0,65	77,6%	78,4%	78,9%	79,6%	80,8%	82,8%	86,2%	92,2%
0,70	80,0%	80,9%	81,3%	82,1%	83,4%	85,4%	88,9%	95,1%
0,80	84,2%	85,0%	85,5%	86,3%	87,6%	89,8%	93,5%	100,0%
1,00	90,0%	90,9%	91,4%	92,3%	93,7%	96,1%	100,0%	
1,20	93,7%	94,6%	95,2%	96,1%	97,6%	100,0%		
1,40	96,0%	97,0%	97,6%	98,5%	100,0%			
1,60	97,5%	98,5%	99,1%	100,0%				
1,80	98,4%	99,4%	100,0%					
2,00	99,0%	100,0%						

Line drawings: summary

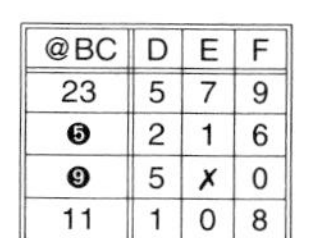

If line drawings are scanned the grey-level limit (brightness) at which the scanner should recognise an area as black is specified. Depending on the original (fine or thick lines, deep black or grey lines) a different value for the limit should be entered. As the 'resolution of the picture' should be much higher than that of the output device, the size of the picture also needs to be taken into account.

In addition, it should be checked whether it is at all sensible to use a scanning resolution which is higher than the physical resolution of the scanner. It is only when scanners are also used to obtain and process grey levels for line drawings that it is worth using an effective scanning resolution of more than 300 or 400 dpi. If a drawing is copied and, if necessary, enlarged before it is scanned, then it is often possible to attain an improvement in the quality of the original picture. The importance of selecting the correct scanning resolution is that photosetting costs and memory requirements may be reduced, without having to accept significant losses in quality.

Formulae

Next, several formulae are given for selecting resolution and image size. The effective scanning resolution should not be higher than the resolution of the output device. The effective scanning resolution is the actual scanning resolution and is identical to the scanning resolution set when the image size is set to 100%.

$$\textit{Scanning resolution} = \textit{desired image resolution} \times \frac{\textit{desired size}}{\textit{size of original}}$$

$$\textit{Image size in \%} = \frac{\textit{desired size}}{\textit{size of original}} \times 100$$

$$\textit{Effektive scanning resolution} = \textit{scanning resolution set} \times \frac{\textit{image size set in \%}}{100}$$

Raster/vector conversion

If raster/vector conversion is carried out memory requirements may be reduced and where applicable the quality of the drawing is increased. The conversion may be carried out either semi-automatically and interactively at the screen or fully automatically. When choosing a raster/vector conversion program care should be taken to ensure that it is simple to operate and may be used to process the drawing further. GEM Artline II, for example, contains functions such as semi-automatic raster/vector conversion, comprehensive manipulations and a variety of drawing capabilities

including colour support and various greyscale progressions. In addition, a 'limit check' is carried out, so that no problems occur when photosetting.

Conversion is usually only worth while if a drawing is to be photoset. Care must therefore be taken that the conversion program can also create files which may be photoset. That is, if a converted curve is too complex the photosetter may fail to process it. For this reason the conversion program must carry out the so-called 'limit check'.

If a raster/vector conversion is to be carried out, scanning should be carried out at a high scanning resolution.

Chapter 13
From PC user to graphics designer

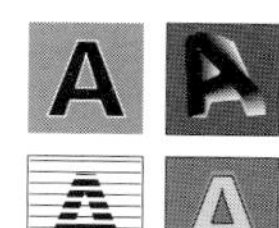

Basic principles of layout

Much is a matter of personal taste. Some experts mock the 'typographic laymen' who, for example, not only underline individual words, but also complete lines.

Much is a matter of personal taste. Some experts mock the 'typographic laymen' who, for example, not only underline individual words, but also complete lines.

Which do you prefer, the first paragraph or the 'correctly' underlined second paragraph? We are of the opinion that you should make use of the modest structuring freedom which DTP offers, even if this contradicts the rules which typography has developed and passed on over many centuries. In the example above, the spaces between words, as well as the descenders, are underlined. This is not permitted according to typographic rules. Nowadays DTP offers all the professional facilities necessary for obtaining an optimum layout, but they are much less complex and also less expensive. DTP is so simple that anyone can learn how to use it.

However, in spite of this, the criticisms from the experts are valid. The problem is not with DTP itself, but with the unrestrained layout possible with it. So many facilities exist, which anyone can use with little effort. The results often reflect this. Anyone who starts working with DTP, but does not have experience of the graphics industry, initially makes the same mistakes. Enthusiasm often leads to every facility being used within a document. Such 'works of art' are not viewed either as good or uncommon from an aesthetic point of view. One reason they are not uncommon is that the number of enthusiastic DTP users is increasing every day. Everything important is made large and bold. And, of course, as everything is important...

After the initial surge of enthusiasm, you should bear in mind that 'less is often more'!

In this book we do not wish to describe in detail how to lay out a document correctly. A complete series of other books are devoted to this subject. However, in spite of this, we would like to point out a few important issues. In principle, you should not

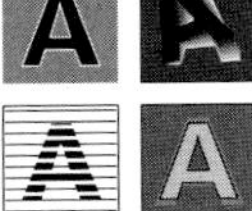

be able to tell whether a DTP product was created using DTP. Therefore, please do not fill the complete page with frames, different raster areas, many lines of different thicknesses, many non-uniform fonts, etc.

Just try it once. Only make text bold if it really has to be bold. Reduce the size of text until you believe that it is too small. Then reduce it considerably more, since from experience the smallest size of text you decide upon initially for the layout is still far too large at the end. You should be frugal in making text stand out and the best way of achieving this is through the use of eye-catching points. More important than large and bold headings is the sensible division of text. The individual blocks of text should not be too small and fragmented, but also not too large.

As the eye cannot focus on too much at once and requires points of orientation, this should be taken into account when arranging the layout. For example, an advertisement may be divided into three blocks: an eye-catcher, an idea of the contents attainable with a single glance, and text which should be very small and specifically there only for those people who wish to know more. Leave sufficient empty space on a page and use paragraphs for longer pieces of text. The eye needs a few less visible orientation and reference points. Subheadings in a longer piece of text are also a good idea. In addition, stick to one, or at most two, different fonts.

We really cannot repeat this often enough: 'less is more'! What is the point of using every scrap of space in a brochure, if the brochure is then no longer readable? The conviction to retain empty space and to create an arrangement which may be grasped with a single glance is important. In addition, make sure that headings are not too large and detailed, that the block of text is small and easy to read and that everything else is split into blocks which are easily surveyed.

Above all, it is important that you leave plenty of blank space between the individual blocks, so that the impression of lightness and air is created. Imagine a page from a newspaper which is completely filled uniformly with text. Even if a subject was of interest, you would not read it as you would not wish to spend so much time on it. However, if the same text was divided with subheadings, pictures and paragraphs you would be quickly enticed to read it.

This means that curiosity may be aroused through an interesting, loose presentation. If longer pieces of text are split into smaller blocks, the feeling that it is not essential to read everything is obtained. This impression reduces the level of resistance which prevents people from beginning to read at all. The courage to include blank space is important in all fields. Since presenting far too much information may result in none at all being taken in, it is better only to include the most important information, so that this at least is taken on board.

At this point we would like to quote the graphics designer and book designer, Jan Tschichold:

 *'Art exists to a large extent in spaciousness'.*

Joy of experimenting is also required

Good design is not based on a flash of inspiration, but usually as a result of many experiments. Even professionals do not usually solve problems at the first attempt. Each trial builds on the previous one and DTP allows you to carry out many experiments in a very short time. Only those people who have the courage always to question their designs will produce really good results.

A technical manual requires a different style to a company newsletter or an advertisement. It is also helpful to orientate yourself on objects which are already available. You should find a good sample document and then try to copy the layout and the style. In this way major errors can be avoided at the start. The same layout scheme should be maintained throughout a document.

Simple methods are available to monitor the layout. View the pages where you have carried out the layout at a distance. If everything seems to merge together and no structure is easily recognisable then something is wrong. An even better method is to permit various people to scrutinise your layout. But be careful: receive any criticisms as positive – a skill that needs to be learned. People fall in love with their own ideas and often easily become 'operation blind'. In addition, look at other publications and try to work out why you like or dislike this or that.

Business graphics

DTP is used more and more often for the creation of business reports, which also include business graphics. Pictures are not only entered using a scanner, but also created through various programs. By using spreadsheet or presentation programs, bar charts, line graphs and pie charts may be created from the numbers entered. A fantastic possibility now exists for changing these standard graphs according to your own personal ideas. Using vector-orientated drawing programs such as GEM Artline, for example, these graphs may be processed further. If the file sizes do not match, then so-called conversion programs may be of use. Many presentation programs also contain drawing functions. Finally, text and pictures are laid out using the DTP program and a professional report is then ready. Your next salary increase should then follow this!

Advertising

Advertising in conjunction with DTP is an important subject, as more and more users are using DTP specially for this purpose. A reason for this, among other benefits, is the speed with which an advertisement, for example, may be created for a newspaper. Advertising, however, is also a particularly difficult subject. Here it becomes especially clear that DTP may be an outstanding technical aid for creating camera-ready copy quickly and inexpensively. But whether or not your advertisement is successful as a result of this is another matter. For good advertisements a proficient layout ability is required and to this end we would like to describe a few principal points.

Not too large, just friendly

Advertisements must fit the situation. For an advertisement in a newspaper speed of recognition is often more important than the technical quality. Often it is sufficient to carry out the printing using a laser printer. However, if brochures are to be created, the quality must naturally be rather higher. Here it is not just the contents which are important, but to a very high degree, the presentation. Even if every aspect of the layout is correct, but the quality of printing leaves a lot to be desired, then this is likely to do more harm than good to your business. Therefore, always include photosetting of your printing files in your plans for large advertising campaigns.

What would the customer like?

What use is the best paper and the highest printer resolution if the layout is not right? It is as difficult as it sounds simple. Consider what the customer wants!

Customers essentially have no time and therefore no desire to read long pieces of text. In addition, they do not initially wish to receive comprehensive and detailed information, but simply wish to review the article at a glance. Hence a few, easily read and readily grasped headlines are important to arouse interest and so prevent the frequent confinement to the bin. Using easily understood small units should then ensure that the customer carries on reading.

The mistake is often made of using too many phrases which the customer has already read or heard x number of times elsewhere, so that they are now worn out. You wish to make a positive impression, but you cannot achieve this if the headlines are too large, with respect to the surrounding material, and hence not very readable. This is also true for the accompanying text. The philosophy should not be 'the bigger, the better', but should instead be one of attaining an easily understood and interesting layout. Therefore, do not make everything so large and massive. However, a certain tension may of course be created using one or two disharmonious elements.

Advertisements

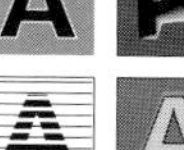

Once again let us make a few more remarks about advertising in a newspaper. Advertisements are more likely to be noticed if sufficient blank space is included and if the advertisement space contains only a single, large and prominent element. The mistake is often made that the complete space should be used fully and so the ratio between blank and filled areas is incorrect. Furthermore, this is true not just for advertisements.

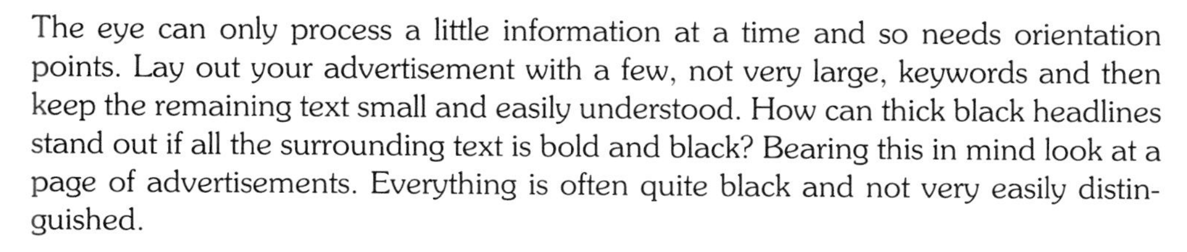

The eye can only process a little information at a time and so needs orientation points. Lay out your advertisement with a few, not very large, keywords and then keep the remaining text small and easily understood. How can thick black headlines stand out if all the surrounding text is bold and black? Bearing this in mind look at a page of advertisements. Everything is often quite black and not very easily distinguished.

If you create a clear, easily distinguished rest point in your advertisement, then you will have already gained a great deal. A generous white area may become the point of attraction in a page of advertisements. Hence it is better to make the headlines a little smaller, use a little less text and use a smaller size text, thereby leaving more space empty.

An important aspect of advertising is that an accompanying picture should not be changed continually. For example, place the your company logo in an advertisement in the same way as it is printed on your letterhead.

Page planning for newspapers

When newspapers are to be printed, so-called page planning (imposition) needs to be carried out. For example, when printing, eight pages may be arranged on a printing plate. These pages later contain printed page numbers depending precisely on the complete product. If one page is to be printed in colour, then all eight pages on the printing plate could be printed in colour without incurring further costs.

For small projects, which are not reproduced by means of printing plates but by copying processes instead, page planning may be carried out automatically. If, for example, a newspaper is to appear at A4 size, then it is made up of individual A3 sheets printed on both sides. As four A4 pages may be fitted on to an A3 sheet, the complete number of pages must of course be divisible by four. For further information refer to the point on 'Imposition' in Chapter 9.

Fonts and dimensions

Text height

Text height determines the size of a letter. For this purpose measurement is made from the top edge of a capital letter (e.g. A) to the bottom edge of a small letter, including any descender (e.g. y). Therefore, it is not only the height of a single letter which is important, but rather the complete height of a line including the descenders for small letters. This text height is specified in points. If this measurement is 10 points, for example, then a 10 point text height is being specified.

Conversion of typographical points into millimetres

Two point systems exist for determining the text height:

- the cicero point system (also known as Didot point system) used in German- and French-speaking countries, and
- the pica point system used in English-speaking countries.

Both point systems are similar, so that it is not very important to differentiate between them. In DTP programs the text height is usually given in the pica point system, so it is advisable to enter the line spacing in partial points. In addition, the text size is not fixed exactly to a specific height, but varies from font to font.

The following formulae are valid for the pica point system:

- Text height in points × 0.351 = text height in millimetres (mm)
- Text height in millimetres/0.351 = text height in points

Explanation of point system

The traditional typographical unit used in German- and French-speaking countries is the 12 partial point (Didot points) divided Cicero system. In English-speaking countries the pica is divided into 12 partial points.

The size of a cicero or pica is very similar:

- 1 pica = 12 partial points = 4.21 mm
- 1 cicero = 12 partial points = 4.51 mm

Converted to a single point this gives the following values:

- 1 pica partial point = 0.351 mm
- 1 cicero partial point = 0.376 mm

The pica partial point thus deviates from the cicero partial point by 0.025 mm, so that 10 point text varies by 0.25 mm. The exact value for a cicero point is 0.376 065 mm. However, as the metric system is being adopted more and more, even by typographers, a value of 0.375 mm was specified for the cicero point (Didot points) on 1 January 1987 for the metric system.

The following examples use pica partial points (English-speaking countries) and the corresponding text size in millimetres (Warning: Both the descenders and ascenders are included in the measurements.):

- 7 point 2,46 mm Consultation size
- 8 point 2,81 mm Consultation size
- 9 point 3,16 mm Reading size
- 10 point 3,51 mm Reading size
- 11 point 3,86 mm Reading size
- 12 point 4,21 mm Reading size
- 14 point 4,91 mm Presentation size
- 16 point 6,62 mm Presentation size
- 18 point 6,32 mm Presentation size
- 20 point 7,02 mm Presentation size
- 24 point 8,42 mm Presentation size
- 28 point 9,83 mm Presentation s
- 32 point 11,23 mm Presenta
- 36 point 12,64 mm Presen
- 40 point 17,55 mm Pre

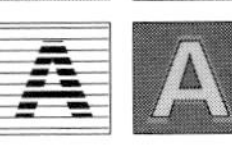

How do we read?

Often text may be found in advertising brochures and advertisements which is too large, used in the hope that it will be noticed and therefore read. This is, however, not so in every case, since a longer piece of text with a height which is too large will not be very clear and is poorly readable. A large, short heading, however, arouses interest and is usually read.

In a long piece of text, where the size of text is too large, the opposite is achieved. It is not easily read and often not even noticed. In order to get a good feel for the text height, we need to be aware of how we read. When we read we do not scan individual letters one by one, but instead a complete picture of several letters and words is viewed simultaneously. In this complete picture only the centre letters are discerned exactly, with the border areas not clearly viewed.

When reading the eye moves along the lines discontinuously in steps of about a centimetre. Therefore, for text with an average height, in one glance about five to ten letters are observed, but only about three letters are focused upon exactly. The others taken notice of are blurred. During the pause between the discontinuous glances only the fixed three letters are completely recognised. However, if these alone do not manage to convey to the reader an idea about what is being read, then the eye must return to make sure of the meaning. Using an optimum text size the meaning is more easily understood and hence fewer returns need to be made.

In this context it is interesting to know that we recognise words mainly by the top half of the letters. Our eye 'photographs', within a short period, pictures of sections of lines one after another. These word pictures are compared with word pictures already stored in our memory. If the 'photograph' matches a word picture in our memory, then the word is recognised.

Text and serifs

However, it is not just the size of text which determines how readable a text is. For longer pieces of text a serif font should always be used. Serifs are small lines of various forms at the periphery of individual letters. A typical example of a serif font is 'Times'. In addition, the font used in this book, i.e. Souvenir, is a serif font. In longer pieces of text it is easier to read a serif font than a font without serifs (e.g. 'Helvetica'), since the serifs emphasise the text.

In general, it is true, not only in connection with serifs, that any passage of text which tires the eye will not be enjoyable to read. A line spacing which is too narrow is just as tiring as lines which are too long, or long pieces of text in an italic font. The same applies if orientation points are missing. The orientation points for the individual letters are the serifs, and those for the text are the paragraphs, indents and subheadings. For this reason, blocked text is also a little less readable than left-justified text.

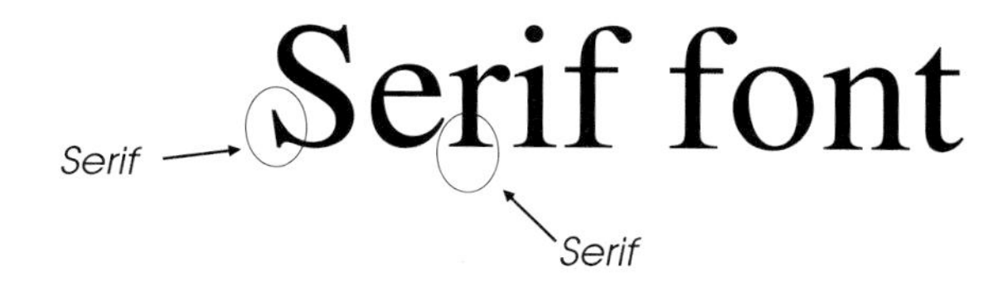

The following shows a few reading examples illustrating the various types of text:

Blocked text using a font without serifs is less readable:
However, it is not just the size of text which determines how readable a text is. For longer pieces of text a serif font should always be used. Serifs are small lines of various forms at the periphery of individual letters. A typical example of a serif font is 'Times'. In addition, the font used in this book, i.e. Souvenir, is a serif font. In longer pieces of text it is easier to read a serif font than a font without serifs (e.g. 'Helvetica'), since the serifs emphasise the text.

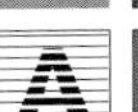

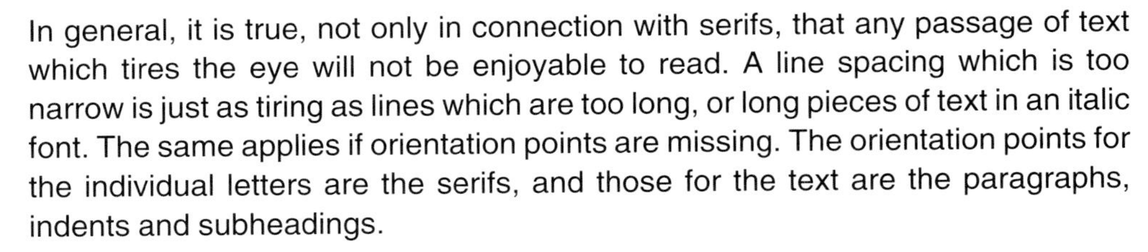

In general, it is true, not only in connection with serifs, that any passage of text which tires the eye will not be enjoyable to read. A line spacing which is too narrow is just as tiring as lines which are too long, or long pieces of text in an italic font. The same applies if orientation points are missing. The orientation points for the individual letters are the serifs, and those for the text are the paragraphs, indents and subheadings.

Longer paragraphs using a blocked font are less readable:
However, it is not just the size of text which determines how readable a text is. For longer pieces of text a serif font should always be used. Serifs are small lines of various forms at the periphery of individual letters. A typical example of a serif font is 'Times'.
In addition, the font used in this book, i.e. Souvenir, is a serif font. In longer pieces of text it is easier to read a serif font than a font without serifs (e.g. 'Helvetica'), since the serifs emphasise the text. In general, it is true, not only in connection with serifs, that any passage of text which tires the eye will not be enjoyable to read. A line spacing which is too narrow is just as tiring as lines which are too long, or long pieces of text in a bold font.

Longer pieces of text in an italic font tire:
However, it is not just the size of text which determines how readable a text is. For longer pieces of text a serif font should always be used. Serifs are small lines of various forms at the periphery of individual letters. A typical example of a serif font is 'Times'. In addition, the font used in this book, i.e. Souvenir, is a serif font. In longer pieces of text it is easier to read a serif font than a font without serifs (e.g. 'Helvetica'), since the serifs emphasise the text. In connection with serifs it is true that any passage of text which tires the eye will not be enjoyable to read.

TEXT WHICH EXCLUSIVELY CONSISTS OF CAPITAL LETTERS IS EXTREMELY UNREADABLE:
HOWEVER, IT IS NOT JUST THE SIZE OF TEXT WHICH DETERMINES HOW READABLE A TEXT IS. FOR LONGER PIECES OF TEXT A SERIF FONT SHOULD ALWAYS BE USED. SERIFS ARE SMALL LINES OF VARIOUS FORMS AT THE PERIPHERY OF INDIVIDUAL LETTERS. A TYPICAL EXAMPLE OF A SERIF FONT IS 'TIMES'. IN ADDITION, THE FONT USED IN THIS BOOK, I.E. SOUVENIR, IS A SERIF FONT. IN LONGER PIECES OF TEXT IT IS EASIER TO READ A SERIF FONT THAN A FONT WITHOUT SERIFS (E.G. 'HELVETICA'), SINCE THE SERIFS EMPHASISE THE TEXT. IN GENERAL, IT IS TRUE, NOT ONLY IN CONNECTION WITH SERIFS, THAT ANY PASSAGE OF TEXT WHICH TIRES THE EYE WILL NOT BE ENJOYABLE TO READ.

Blocked text or left justified?

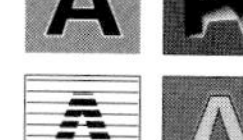

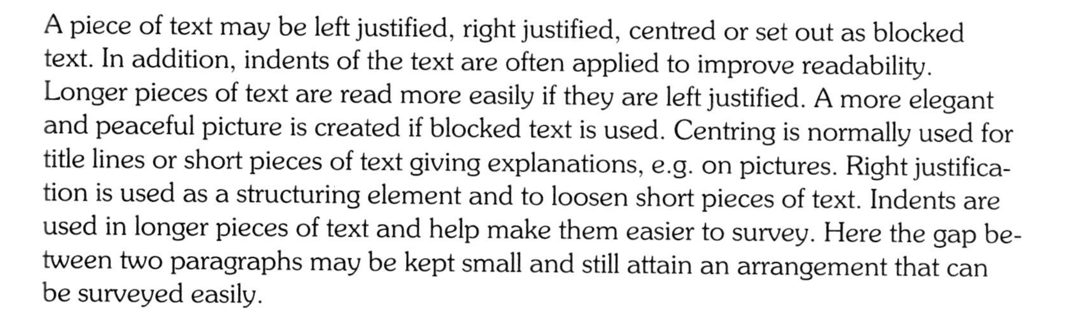

A piece of text may be left justified, right justified, centred or set out as blocked text. In addition, indents of the text are often applied to improve readability. Longer pieces of text are read more easily if they are left justified. A more elegant and peaceful picture is created if blocked text is used. Centring is normally used for title lines or short pieces of text giving explanations, e.g. on pictures. Right justification is used as a structuring element and to loosen short pieces of text. Indents are used in longer pieces of text and help make them easier to survey. Here the gap between two paragraphs may be kept small and still attain an arrangement that can be surveyed easily.

Letter widths and word spacing

As the letter widths vary in fonts which have proportional spacing, a so-called letter width table is required in which the individual letter widths are given. Word spacing (gap between words) should in general be one-third of the height of the text, i.e. in a 10 point text the word spacing is about 3.3 points wide.

Leading (line spacing)

Text height plus leading equals the line spacing. The leading is the gap between the lower edge of a small letter with descender (e.g. y) and the upper edge of a capital letter (e.g. A) in the next line. The diagram on the next page clarifies these relationships.

Leading is always specified in combination with the text height. If a leading of '10 on 12 point' is specified, this means that for a 10 point text the leading is 2 points. A correctly chosen leading (line spacing) can increase the readability of a piece of text. If the leading is made large enough, the text appears lighter with more air. In this way, especially with long pieces of text, the readiness to read it may be increased.

A rule of thumb is that the line spacing should be 1.2 times the height of the text.

Accordingly, a text height of 10 points has a line spacing of 12 points and hence the leading is 2 points.

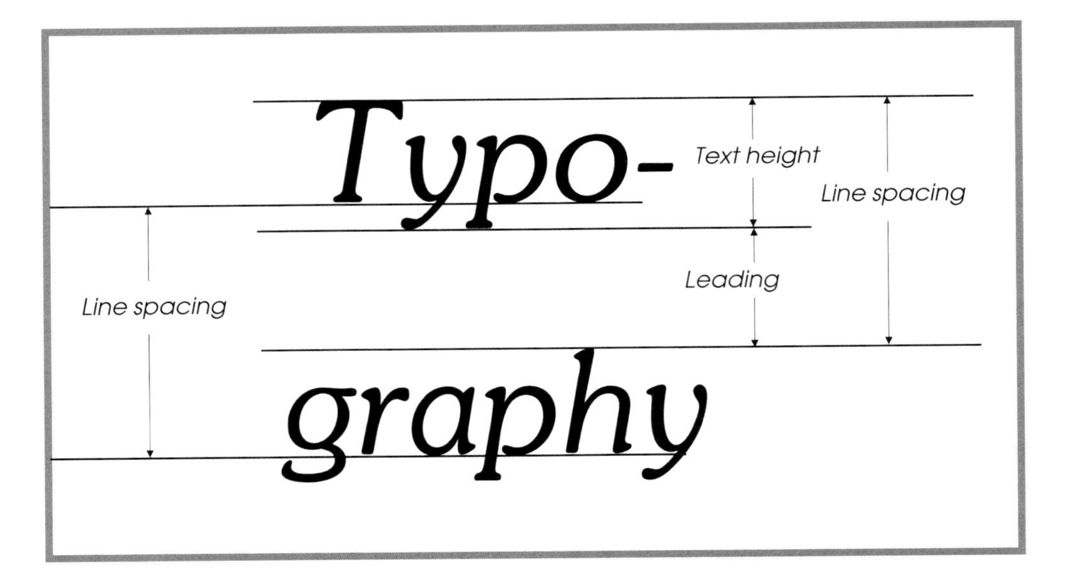

The ease of finding the next line is an important criterion when reading. If you reach the end of a line, you must be able to find the beginning of the next line very quickly without the need to search for it. For this reason a line should not have more than 60–70 characters, otherwise the eye will need to return over a distance which is too long and successive lines may then be easily confused. A text which is to be read quickly and easily should not have more than 35–40 letters per line. However, when finding the next line of text, it is not only the length of the lines that is important, but also the line spacing.

In order to clarify the line spacing problem, two texts are shown below both with the same text height but different leadings:

5 point leading – 10 on 15 point Times

However, it is not just the size of text which determines how readable a text is. For longer pieces of text a serif font should always be used. Serifs are small lines of various forms at the periphery of individual letters. A typical example of a serif font is 'Times'. In addition, the font used in this book, i.e. Souvenir, is a serif font. In longer pieces of text it is easier to read a serif font than a font without serifs (e.g. 'Helvetica'), since the serifs emphasise the text. In general, it is true, not only in connection with serifs, that any passage of text which tires the eye will not be enjoyable to read. A line spacing which is too narrow is just as tiring as lines which are too long, or long pieces of text in an italic font.

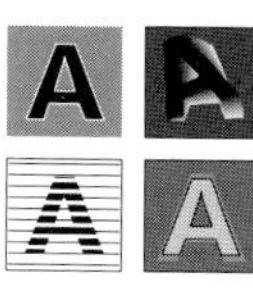

2 point leading – 10 on 12 point Times

However, it is not just the size of text which determines how readable a text is. For longer pieces of text a serif font should always be used. Serifs are small lines of various forms at the periphery of individual letters. A typical example of a serif font is 'Times'. In addition, the font used in this book, i.e. Souvenir, is a serif font. In longer pieces of text it is easier to read a serif font than a font without serifs (e.g. 'Helvetica'), since the serifs emphasise the text. In general, it is true, not only in connection with serifs, that any passage of text which tires the eye will not be enjoyable to read. A line spacing which is too narrow is just as tiring as lines which are too long, or long pieces of text in an italic font. The same applies if orientation points are missing. The orientation points for the text are the paragraphs, indents and subheadings.

Kerning

When letters are pushed together so that text is more readable or looks better, this is known as kerning. It is also known as a change in the letter spacing.

Some letter pairings lead to gaps becoming too small or leave too much white space. Words only appear even and look good if the spacing of the individual letters is not linear, but matched to the space available. This function is used mainly when structuring large headings and titles. For blocked text the processing takes too long and the improvement in appearance is hardly noticeable. Kerning is therefore rarely carried out in such text. Kerning is not related to proportional spacing, where each individual letter has a different width in any case. In kerning, selected letter pairs are pushed a little closer together. The simplest way of explaining kerning is by the following example:

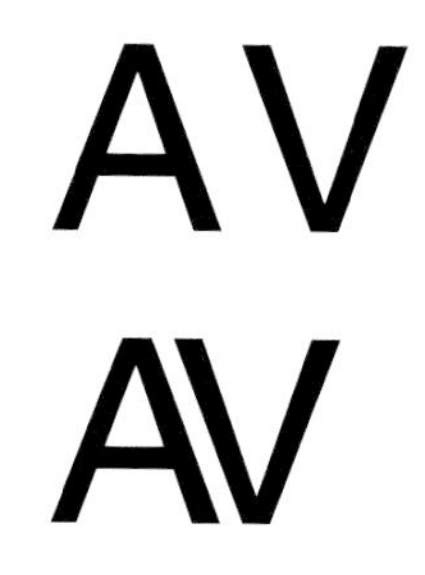

The first letter pair was set normally (without kerning).

The second letter pair was manually kerned.

Kerning is mainly used for large titles.

Widows and orphans

Widows and orphans are the names given to single lines which belong to a paragraph, but due to the page break are isolated either at the end or at the beginning of a separate page. Using a professional DTP program it is possible to specify how many lines of a paragraph are allowed to be isolated at the beginning or end of a page.

'Artistically distorted' text

Using vector-orientated drawing programs it is possible not only to create drawings, but also by simple means to design artistic text creations. Thus, texts which are already available may be transformed as desired.

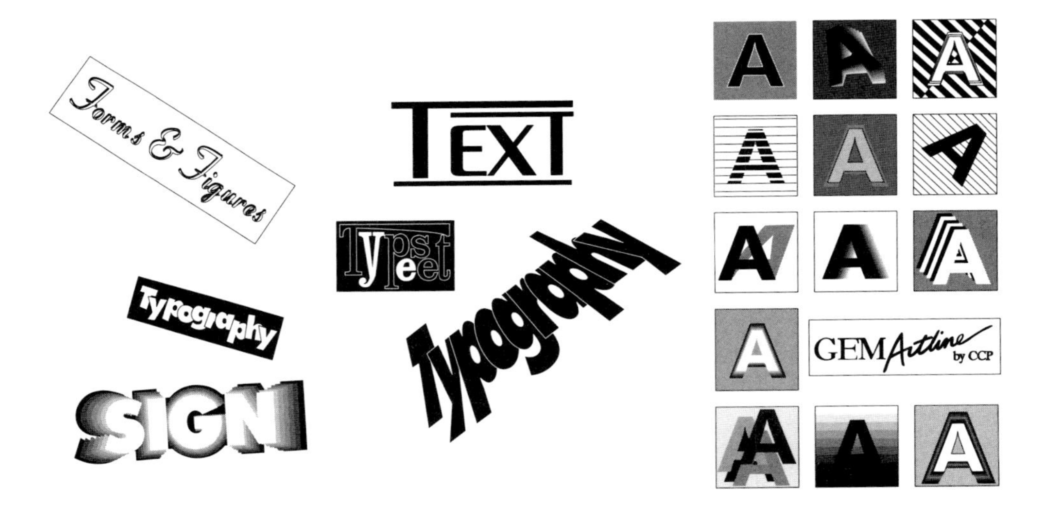

Spaces and quads

A quad is a space of a certain defined size. This size, however, is not specified as a dimension in millimetres, but is given as the width of a letter for the font and text height used.

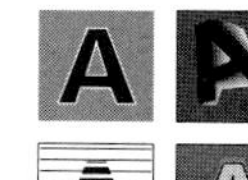

When laying out pages it should be noted that a proportionally spaced text is normally used, i.e. the individual characters are of different widths. This also means, for example, that text indents and tables may not be made using 'spaces'. Outstanding capabilities are included in DTP programs for creating indents and tables, so that quads are rarely required. A standard formatting facility for example is the tabulation of various forms (decimal, right or left justified). The professional version of Ventura Publisher offers the additional capability of automatically creating tables with the relevant frames.

For defined spaces the following are available:

- Quad space = width of the letter M or the symbol @
- Semi-quad space = space of the letter n

In Ventura Publisher, for example, the following exist:

- Small space = width of a dot
- Number space = width of a number

Character set and ASCII table

The individual characters are not stored graphically as pixels, but instead an agreed code is assigned to each character. If, for example, the decimal code 65 is sent to a printer, the printer knows that it should print an 'A'. As the program and the printer need to use the same letter for the code 65, agreement needs to be reached here. One such international agreement is the so-called ASCII code or character set. ASCII stands for '**A**merican **S**tandard Code for **I**nformation **I**nterchange'. Other standards have been developed, based on the ASCII code, but which offer additional capabilities.

Special characters which are not available on the keyboard may be entered readily. In personal computers you press the ALT key and at the same time use the numeric key pad on the right hand side of the keyboard to enter the numeric code for the particular character. For example, the numeric code 1, 2, 3 entered with the ALT key pressed down produces a left curly bracket ({).

Depending on the printer used the characters and the codes assigned to them may change. These special characters are sometimes not displayed on the screen but only appear when printed. It is possible to obtain additional character sets to extend the capabilities of a printer for the most diverse purposes.

However, in Windows programs it is not the ASCII code which is entered, but the ANSI code instead. This character set has been developed by the American National Standards Institute (ANSI).

Below is a list of the PostScript character sets for Ventura Publisher. Characters 32 to 126 inclusive correspond to the ASCII character set. Which code should be used for a particular character may be deduced from the relevant manual or by means of a test print.

Decimal	ANSI		Standard	Symbol	Dingbats
33	033	=	!	!	✁
34	034	=	"	∀	✂
35	035	=	#	#	✃
36	036	=	$	∃	✄
37	037	=	%	%	☎
38	038	=	&	&	✆
39	039	=	’	∋	✇
40	040	=	(	(	✈
41	041	=	)	)	✉
42	042	=	*	∗	☛
43	043	=	+	+	☞
44	044	=	,	,	✌
45	045	=	-	−	✍
46	046	=	.	.	✎
47	047	=	/	/	✏
48	048	=	0	0	✐
49	049	=	1	1	✑
50	050	=	2	2	✒
51	051	=	3	3	✓
52	052	=	4	4	✔

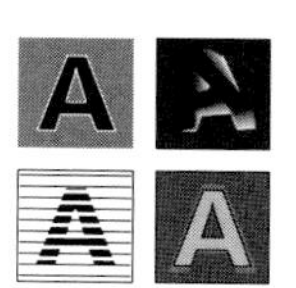

Decimal	ANSI		Standard	Symbol	Dingbats
53	053	=	5	5	✕
54	054	=	6	6	✖
55	055	=	7	7	✗
56	056	=	8	8	✘
57	057	=	9	9	✙
58	058	=	:	:	✚
59	059	=	;	;	✛
60	060	=	<	<	✜
61	061	=	=	=	✝
62	062	=	>	>	✞
63	063	=	?	?	✟
64	064	=	@	≅	✠
65	065	=	A	Α	✡
66	066	=	B	Β	✢
67	067	=	C	Χ	✣
68	068	=	D	Δ	✤
69	069	=	E	Ε	✥
70	070	=	F	Φ	✦
71	071	=	G	Γ	✧
72	072	=	H	Η	★
73	073	=	I	Ι	✩
74	074	=	J	ϑ	✪
75	075	=	K	Κ	✫
76	076	=	L	Λ	✬
77	077	=	M	Μ	✭
78	078	=	N	Ν	✮
79	079	=	O	Ο	✯

Decimal	ANSI		Standard	Symbol	Dingbats
80	080	=	P	Π	✰
81	081	=	Q	Θ	✱
82	082	=	R	Ρ	✲
83	083	=	S	Σ	✳
84	084	=	T	Τ	✴
85	085	=	U	Υ	✵
86	086	=	V	ς	✶
87	087	=	W	Ω	✷
88	088	=	X	Ξ	✸
89	089	=	Y	Ψ	✹
90	090	=	Z	Ζ	✺
91	091	=	[	[	✻
92	092	=	\	∴	✼
93	093	=	]	]	✽
94	094	=	^	⊥	✾
95	095	=	_	‾_	✿
96	096	=	‘		❀
97	097	=	a	α	❁
98	098	=	b	β	❂
99	099	=	c	χ	❃
100	0100	=	d	δ	❄
101	0101	=	e	ε	❅
102	0102	=	f	φ	❆
103	0103	=	g	γ	❇
104	0104	=	h	η	❈

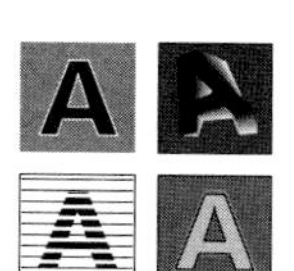

Decimal	ANSI		Standard	Symbol	Dingbats
105	0105	=	i	ι	❉
106	0106	=	j	φ	❊
107	0107	=	k	κ	❋
108	0108	=	l	λ	●
109	0109	=	m	μ	❍
110	0110	=	n	ν	■
111	0111	=	o	ο	❏
112	0112	=	p	π	❐
113	0113	=	q	θ	❑
114	0114	=	r	ρ	❒
115	0115	=	s	σ	▲
116	0116	=	t	τ	▼
117	0117	=	u	υ	◆
118	0118	=	v	ϖ	❖
119	0119	=	w	ω	◗
120	0120	=	x	ξ	❘
121	0121	=	y	ψ	❙
122	0122	=	z	ζ	❚
123	0123	=	{	{	❛
124	0124	=	|	|	❜
125	0125	=	}	}	❝
126	0126	=	~	∼	❞
127		=		✗	
128	0199	=	Ç		
129	0252	=	ü	ϒ	❡

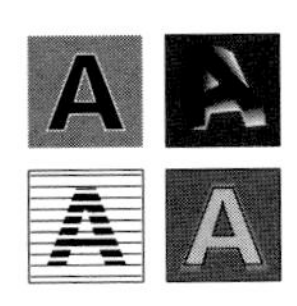

Decimal	ANSI		Standard	Symbol	Dingbats
130	0233	=	é	′	❢
131	0226	=	â	≤	❣
132	0228	=	ä	⁄	❤
133	0224	=	à	∞	❥
134	0229	=	å	ƒ	❦
135	0231	=	ç	♣	❧
136	0234	=	ê	♦	♣
137	0235	=	ë	♥	♦
138	0232	=	è	♠	♥
139	0239	=	ï	↔	♠
140	0238	=	î	←	①
141	0236	=	ì	↑	②
142	0196	=	Ä	→	③
143	0197	=	Å	↓	④
144	0201	=	É	°	⑤
145	0230	=	æ	±	⑥
146	0198	=	Æ	″	⑦
147	0244	=	ô	≥	⑧
148	0246	=	ö	×	⑨
149	0242	=	ò	∝	⑩
150	0251	=	û	∂	❶
151	0249	=	ù	•	❷
152	0255	=	ÿ	÷	❸
153	0214	=	Ö	≠	❹
154	0220	=	Ü	≡	❺

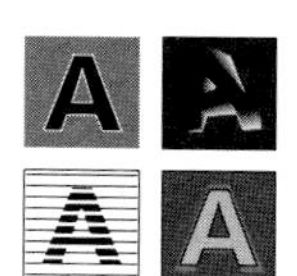

Decimal	ANSI		Standard	Symbol	Dingbats
155	0162	=	¢	≈	❻
156	0163	=	£	…	❼
157	0165	=	¥	|	❽
158	0164	=	¤	—	❾
159	0136	=	ƒ	↵	❿
160	0225	=	á	ℵ	①
161	0237	=	í	ℑ	②
162	0243	=	ó	ℜ	③
163	0250	=	ú	℘	④
164	0241	=	ñ	⊗	⑤
165	0209	=	Ñ	⊕	⑥
166	0170	=	ª	∅	⑦
167	0186	=	º	∩	⑧
168	0191	=	¿	∪	⑨
169	0147	=	“	⊃	⑩
170	0148	=	”	⊇	❶
171		=	‹	⊄	❷
172		=	›	⊂	❸
173	0161	=	¡	⊆	❹
174	0171	=	«	∈	❺
175	0187	=	»	∉	❻
176	0227	=	ã	∠	❼
177	0245	=	õ	∇	❽
178	0216	=	Ø	®	❾
179	0248	=	ø	©	❿

Decimal	ANSI		Standard	Symbol	Dingbats
180		=	œ	™	➔
181		=	Œ	Π	→
182	0192	=	À	√	↔
183	0195	=	Ã	·	↕
184	0213	=	Õ	¬	➘
185	0167	=	§	∧	➙
186		=	‡	∨	➚
187		=	†	⇔	➛
188	0182	=	¶	⇐	➜
189	0169	=	©	⇑	➝
190	0174	=	®	⇒	➞
191	0153	=	™	⇓	➟
192	0132	=	„	◊	➠
193	0133	=	…	〈	➡
194		=	‰	®	➢
195		=	•	©	➣
196	0150	=	–	™	➤
197	0151	=	—	Σ	➥
198	0176	=	°	⎛	➦
199	0193	=	Á	⎜	➧
200	0194	=	Â	⎝	➨
201	0200	=	È	⎡	➩
202	0202	=	Ê	⎢	➪
203	0203	=	Ë	⎣	➫
204	0204	=	Ì	⎧	➬

Decimal	ANSI		Standard	Symbol	Dingbats
205	0205	=	Í	⎨	➭
206	0206	=	Î	⎩	➮
207	0207	=	Ï	⎪	➯
208	0210	=	Ò		
209	0211	=	Ó	〉	➱
210	0212	=	Ô	∫	➲
211		=	Š	⌠	➳
212		=	š	⎮	➴
213	0217	=	Ù	⌡	➵
214	0218	=	Ú	⎞	➶
215	0219	=	Û	⎟	➷
216		=	Ÿ	⎠	➸
217	0223	=	ß	⎤	➹
218		=	Ž	⎥	➺
219		=		⎦	➻
220		=		⎫	➼
221		=		⎬	➽
222		=		⎭	➾

Chapter 14
A DTP workstation

Which requirements must a DTP system meet?

In all cases you need a fast computer with a processor speed of at least 33 MHz, a hard disk with at least 150 Mbytes, etc.... No, we do not want to begin like this.

Often a customer is completely isolated when it comes to setting up an optimum DTP workstation. All too frequently one-sided reference is made of superficial technical performance. However, it is important to consider all the factors and an important one here is the user. The user should be able to entice creative and high-quality performance from the machine. Why is it then so rare that mention is made of the importance of the user? Is it because it is too expensive? On the contrary, it becomes expensive if the importance of the user is not considered. Labour is expensive and motivation cannot be enforced. In addition, it is often 'know-how' which is the key to success.

Example: You buy a DTP system with a 20" double-page monitor. In the PC dealer's showroom you were impressed with the representation and high resolution of the monitor. Now, after you have started working with it, you encounter undefinable

Working effectively: ergonomics is an important aspect when setting up a workstation.

problems: headaches, dizziness, watering eyes or tiredness. Would you think that the problem could be solved by your monitor which impressed you so much? Certain technical data is often quoted incorrectly, such as the refresh rate of the monitor and the graphics card. If the refresh rate is too low, the monitor flickers almost unnoticed. Even if you do not notice it, your eyes still register this flicker with the consequences mentioned above.

However, even if the refresh rate is specified it is often not achieved in practice. The frequency of the monitor is of little importance if the associated graphics card is less powerful. Often the graphics card has either a good resolution or an acceptable refresh rate, but not both at the same time! If you install the incorrect drivers, you will work with an outstanding monitor at a high resolution, but with a refresh rate that is much too low. This results in headaches, tiredness, etc. You can find out more about this subject at the end of this chapter under 'Ergonomic viewpoints on the workstation'.

Next, we meet the two worlds: MS-DOS with IBM-compatible personal computers and Apple Macintosh. Apple computers currently have a 10% share of the market. The software for Apple computers cannot be used on MS-DOS personal computers and vice versa. This emphasises the dilemma. However, it is not just the operating system but also the hardware which must recognise and support the programs used. Personal computers form a large family, with the most important functions being identical. They vary only in detail such as the speed and size of the hard disk, or in a few additional functions and working parts. As mentioned, this has the advantage that the same program may be run on a personal computer from company X in the same way as it may be run on a personal computer from company Y. As a market exists for millions of the same type of computer, it is beneficial for software developers to write comprehensive programs.

Therefore, if you wish to have access to the largest software library in the world, you cannot ignore MS-DOS personal computers. However, this does not yet mean that an MS-DOS personal computer is the only system which should ever be considered, as other computer systems also have advantages in certain fields. It may occur, for example, that you find certain very specific details important which are not (yet) available in the MS-DOS world. In the beginning DTP was carried out mainly on Apple computers. Even now the Macintosh certainly has some advantages. However, which computer world you decide upon in the end depends also on what equipment you have already and what you wish to do besides DTP. The decision certainly depends on whether all your work is purely graphical or whether DTP is just one task among many. Nowadays the Apple Macintosh and MS-DOS personal computer worlds are certainly considered as equals in the DTP field.

Monitors and graphics cards for DTP

The following details relate primarily to MS-DOS personal computers. However, you will also meet other systems in the key technical statements.

A monitor and the associated graphics card should have at least the following characteristics for a DTP system:

- Size

  Monitor, 20" or larger.

- Resolution

  At least 1024 × 768 dots.

- Refresh rate

  At least 75 Hz, non-interlaced (without line jumps).

  A refresh rate of at least 70 Hz is a requirement for all 14" screens nowadays, while good monitors already attain frequencies of over 80 Hz. As flickering is more noticeable in larger monitors, the refresh rate for these should be at least 75 Hz.

- Type of representation

  Black text on white background.

- Colour

  The selection of a colour monitor depends on the application and above all on financial means. If necessary, a secondary colour VGA monitor may be used. For professional colour applications a monitor is required, including graphics card, which is capable of representing the colours red, green and blue each in at least 64 colour levels. By mixing these three primary colours about 262 000 different colours can be attained (64 × 64 × 64 = 262 144). However, it is already possible to obtain realistic photographic pictures with only 32 000 colours.

- Grey levels

  At least 64 grey levels with a resolution of at least 640 × 400 dots.

  The representation of grey levels is required in order to evaluate greyscale pictures. However, even without a greyscale monitor pictures may still be represented, but in this case they are then rasterised. Therefore, it is not possible to carry out image processing.

  It is also possible to work with a two-screen system: a 20" double-page monitor and a VGA monitor are operated at the same time by the computer system. It is then possible to switch between the two screens. Here care must be taken that a VGA graphics card is used which offers 64 grey levels and a resolution of at least 640 × 400 dots. The programs used must support the VGA card.

By the way, standard VGA graphics cards are able to represent 256 colours from a palette of 262 000 colours, but are only capable of representing 64 grey levels. Using special graphics cards and monitors it is also possible to represent 256 grey levels. However, for image processing a representation using 64 grey levels is sufficient. But for professional applications the monitor should be capable of representing 256 grey levels, primarily in order to be able to evaluate the dark areas of a picture sufficiently.

- Miscellaneous

  The monitor should be anti-dazzle. It is possible to obtain certain monitors which are not anti-dazzle. Anti-dazzle is obtainable at a price, but in this case the money is certainly invested well. Therefore, when buying the monitor make sure that it is anti-dazzle.

- Drivers for the current software

  The best monitor and the best graphics card are worthless if so-called 'drivers' do not exist to drive them. This means that, to drive the graphics card, a software driver is required embedded within the application program, so that the facilities of the graphics card may be used. In MS-DOS computers, processing is usually carried out under Windows, so that a driver written for Windows is sufficient.

Monitors and graphics cards for word processing

As the subject of monitors is so important, we would also like to describe the requirements profile of a monitor for word processing. For this a good monitor and its associated graphics card should have the following characteristics:

- Size and resolution

  14", at least VGA standard.

- Refresh rate

  At least 70 Hz (80 Hz is better), non-interlaced.

  Only when a monochrome monitor with amber or green colours is installed may a refresh rate of less than 70 Hz be used, since these monitors have an illuminated layer enabling the letters to stay illuminated longer. However, you should not switch to inverse video representation here.

- Type of representation

  Black text on white background with overscan technology (no black border). If the characters are not represented as black on a white background, then problems associated with health may occur. This is particularly the case if you are continually changing your attention from a light original document on a table to a dark monitor and vice versa. This does not apply to most programmers, since they are looking almost continuously at the same monitor. It mainly applies to those people who have to type whilst copying from an original document.

- Colour

  Monochrome monitor.

  A standard colour monitor is not usually suitable for word processing. A refresh rate of 70 Hz is not attained and the characters are not displayed as black on a white background. Standard VGA graphics cards only attain a refresh rate of 60 Hz in VGA graphics mode. Other specifications are often heard which are incorrect. A standard VGA graphics card only attains a refresh rate of 70 Hz in EGA and text mode. An exception here is the range of special graphics cards and colour monitors which attain refresh rates of 70 Hz, or even more than 90 Hz. However, in this case a special monitor or multiscan monitor needs to be used.

- Note

  With a 20" double-page monitor word processing can become a special experience, since you are able to see a complete page at a glance. A double-page monitor also has the advantage that long lines are displayed entirely on the screen.

The computer system

Anyone who wishes to work professionally with DTP requires a 'fast' computer. Graphics applications require a large amount of time and memory. Anyone who tries make savings here will certainly be annoyed with themselves at a later stage. However, be careful. It is not only keywords such as '33 MHz' which are important.

An important criterion is the speed of the hard disk (e.g. less than 20 ms average access time) as well as that of the hard disk controller (effective transfer rate greater than 8 Mbit/s). What point is there in having the fastest computer in the world, if the hard disk turns out to be a bottleneck? Therefore, a fast hard disk is important, especially for DTP applications.

However, the capacity of the storage device is also important. If you only use text processing, then an 80 Mbyte hard disk is more than sufficient. On the other hand, with DTP applications 80 Mbyte hard disks are the minimum requirement. If you work professionally and wish to integrate pictures, then you should start with a disk of at least 150 Mbytes (slimline). At least 4 Mbytes and preferably 8 Mbytes or more should be available as the main memory.

Data storage and data transfer are important issues in DTP applications. It is not possible to store everything on hard disk, since even a 760 Mbyte hard disk would be full within a short time. However, it is also not always possible to delete everything which is no longer currently required.

The following additional storage devices are available for DTP applications:

- 40, 90 or more than 200 Mbyte cartridge units
- 44 Mbyte Bernoulli diskettes
- 20 Mbyte 3.5" diskettes
- 40, 60, 150 or 250 Mbyte magnetic tapes (streamers)
- 600 Mbyte, 2.2 Gbyte DAT tapes (Digital Audio Tape)
- 600 Mbyte, 1 Gbyte optical data storage devices

Anyone who wishes to use DTP professionally needs a powerful computer system.

Photograph: Tandon GmbH

If the interchangeable data storage device is used not only for the transfer of data and data storage, but also as a storage device for current work, then care must be taken that it has a fast access time. It is also important to have a high data transfer rate for data storage.

With streamers individual files are not accessed easily and this takes time. However, streamers do exist which are very reasonably priced, easy to operate and which have small, handy magnetic tape cassettes, so that these devices can certainly be recommended too. However, cartridge units and diskettes, such as Syquest cartridge units, for example, are rather more suitable.

Another minor issue is the mouse. Do not make savings here as the mouse is the most important input device of the DTP system. Above all, it should feel comfortable under the hand. Dynamic mice are preferable with double-page monitors. The distances on the large screen may then be traversed with shorter mouse movements, yet it is still possible to locate objects precisely.

The printer

Issues again become much simpler when considering the subject of the printer. If you wish to use DTP professionally, there is no way of doing so without a PostScript laser printer. However, even here variations exist. For example, the speed of the PostScript interpreter is important. The printer contains a computer, which processes all the PostScript commands. This computer should be as powerful as possible so that you do not have to wait too long for pictures to be printed. (Please do not confuse the performance of the PostScript interpreter with the speed of the printer, which the manufacturer specifies only for the printing mechanism.)

Another advantage of PostScript printers is that text may be represented at any chosen size. A good PostScript printer can even emulate other laser printers such as the HP LaserJet (for example), so that the advantages of these printers do not need to be relinquished. Perhaps you should also consider acquiring a printer with two paper trays (e.g. one tray for letterheaded paper and one tray for ordinary blank paper or envelopes). Of course, you can also obtain good results without a PostScript printer. Naturally you would then be limited in certain ways. For example, the interplay with a photosetter would be rather more difficult (e.g. correct line and page breaks). However, programs are available which translate PostScript commands into the language of the HP LaserJet. Therefore, a PostScript file may be printed even on a non-PostScript printer. The term 'PostScript printer' implies essentially that the printer understands the language 'PostScript'.

Variations in printers also exist due to their drive mechanisms. Standard printers such as impact pin printers, laser printers and ink jet printers are certainly generally well known and do not need to be described in more detail here. In addition, there are also so-called 'sublimation printers' and 'video printers'. These printers can output both real grey levels and colours and so achieve an outstanding quality in pictures. However, even special ink jet printers are capable of outputting greyscale pictures. Ink jet printers are able to create almost 64 different dot sizes and hence 64 grey levels. They are even capable of doing this in colour. The disadvantage, however, is that their resolution of about 160 dpi is relatively low for the representation of text.

Laser printers conforming to different standards are available in various price ranges. Standard laser printers (300 dpi) may produce text of good quality, but the quality of greyscale pictures is not very convincing. Many of these printers are able to improve the sharpness of edges by varying the size of pixels slightly and so produce better quality text. For example, for HP printers this technology is called Resolution Enhanced Technology (RET). Laser printers using DMV technology (Dot Multi Variation) obtain a resolution of approx. 2400 dpi and are relatively inexpense. For data to be photoset a PostScipt laser printer is required for proof prints. This ensures that the layout of the document printed with the laser printer is identical to the layout of the film. However, standard PostScript laser printers with high resolutions are quite expensive.

The scanner

It is not so easy to answer the question of which scanner is the 'best', since an ideal multi-purpose scanner does not yet exist. Choice varies depending on the application and the finance available. Anyone who is looking for an inexpensive scanner will certainly be satisfied with a 300 dpi scanner with 64 or 256 grey levels. In any case, those who print using a laser printer should only use a scanner with a resolution of more than 300 dpi when enlargements are to be made.

However, the decision is rather more difficult for professionals. Issues such as distortion and lines in the picture then need to be taken into account as well as the capabilities of the software. In the end the decision on which scanner is most suitable can only be made after carrying out trials, as the data sheets from the manufacturers do not yet provide sufficient specifications on their scanners. Unfortunately, DTP scanners today only fulfil the minimum requirements, so that time-consuming image processing still needs to be carried out by the software.

Care must be taken to ensure that the scanner resolutions given below are interpreted with respect to the physical or optical resolution and not with respect to the output resolution as mentioned in many brochures. In addition, the number of grey levels must be interpreted with respect to those read by the hardware and not those calculated by interpolation.

Below is a rough classification of scanners. As well as considering the specifications given you must decide whether you require a black and white or colour scanner. You should also pay special attention to the light source used in the scanner in order to determine the eventual implications of drop-out colours.

- Flat-bed scanner: 300 dpi, 64 grey levels

  Even though you can obtain good-quality pictures using scanners which only have 64 grey levels, these scanners are not state of the art. Therefore, when buying a new scanner it should be one with 256 real grey levels. On the other hand, if you already have a scanner with 64 grey levels, you should certainly be satisfied with the attainable picture quality.

- Flat-bed scanner: 300 dpi, 256 grey levels

  Pictures of outstanding quality may be obtained with these scanners. The quality of line drawings is, however, only satisfactory if the original drawings are sufficiently large, so that the pictures can be reduced afterwards, or if they are to be printed only on a laser printer.

  Using suitable software, the quality of line drawings may be improved also by means of interpolation. When scanning greyscale pictures the resolution may be too low if the pictures are to be significantly enlarged. However, it is also possible to carry out enlargements with the 'interpolation' function from an image processing program. The low resolution then has advantages when photographs of standard size are to be scanned and printed.

- Flat-bed scanner: 400 dpi, 256 grey levels

  The physical resolution of 400 dpi is a good starting point for standard applications. Using this resolution greyscale pictures may be significantly enlarged. Also, by using text recognition programs (OCR) better results are attainable. Furthermore, the resolution of 400 dpi is sufficient in most cases for line drawings, particularly if the software allows a higher output resolution to be calculated.

  In those cases where scanning is performed using the optical resolution (e.g. 400 dpi) with data reduction then carried out, large volumes of data are produced due to the high scanner resolution and these need to be processed. A 300 dpi scanner therefore certainly has advantages here over a 400 dpi scanner for standard originals and standard image sizes.

- Flat-bed scanner with more than 400 dpi resolution

  Scanners with more than 400 dpi resolution are only required for line drawings or for pictures which need to be greatly enlarged, for example the scanning of slides.

  An optical resolution of more than 400 dpi is only suitable for the scanning of photographs if the scanner carries out an internal interpolation. That is, if you set a resolution of 200 dpi in a 600 dpi scanner (which does not interpolate) then only every third CCD sensor will be activated, causing the picture to be read in with accompanying 'holes'. This results in poor-quality pictures, since in many cases a scanning resolution of less than 200 dpi is sufficient. Scanners do exist which carry out internal interpolation, but since it is often only carried out line by line, the quality is also diminished.

  Our opinion: considering the current state of the art real 600 dpi scanners are not to be recommended. In any case, some so-called 600 dpi scanners only have an optical resolution of 300 dpi, since only the line feed has a resolution of 600 dpi. These scanners are more suitable for image processing. However, as long as manufacturers continue to publish data sheets with insufficient information and where these issues are not mentioned, you should be careful when buying a 600 dpi scanner.

- Scanners with more than 256 grey levels

  Here, 8 bit scanners read 256, 10 bit scanners read 1924 and 12 bit scanners read 4096 grey levels. Our dream is to have a very fast scanner using 400 dpi optical resolution and 4096 grey levels. With these scanners dark original pictures could be made lighter without any major losses in quality. Using loadable gradation curves an optimum reduction to 256 grey levels could be attained. A reduction to 256 grey levels is paramount as all image processing programs, including even PostScript, are only able to process 256 grey levels.

Flat bed scanners
Photograph: EPSON (UK)

If only scanner manufacturers would admit that their scanners have an optical resolution of only 300 dpi instead of the specified resolution of 600 dpi, then these would be optimum scanners for the reproduction of photographs. The fact that some scanners have a built-in step motor with a higher resolution is not a disadvantage in this context. The resolution of the optics is always important. Unfortunately, newer scanners actually have an optical resolution of 600 dpi, but do not yet interpolate and so are less suitable for picture processing.

However, software should be available for 10 and 12 bit scanners which is capable of utilising the special capabilities of these scanners. It would be better if the basic functions of image processing were already built into the scanner as it could then be operated using standard software.

- Scanners with logarithmic amplifiers

  Scanners which have a logarithmic amplifier built in on the analogue side make the picture lighter before digitising. Hence, the grey levels need only be changed subsequently by a small amount. The 256 grey levels obtained are distributed to an optimum degree and so it is easier to make dark original pictures lighter.

- Camera scanners

  The design of camera scanners is considerably different to that of flat-bed scanners. Fixed on a tripod is a camera in which various objective lenses may be placed. However, CCD sensors are used instead of film in the camera. The original being scanned is illuminated by several light sources. As a result of this design the resolution may be increased to almost any value. By reducing the distance between the camera and the original, the area being scanned becomes smaller and so the resolution becomes correspondingly higher. These scanners therefore attain a resolution of 1000 dpi for A5 originals and even 600 dpi for A4 originals. Even originals of size A0 may still be read at a resolution of 150 dpi.

  The strengths of these scanners are only utilised if very small and very large originals, slides, negatives and very fine line drawings are to be scanned. These advantages, however, must be offset against the high price, complex operation and larger amount of machine space required. Even the calibration is a little more difficult. In general, it is clear that more time must be spent mastering the technology and operation of these scanners than with flat-bed scanners. Mention must also be made of the fact that although inexpensive camera scanners do exist, they have their limitations regarding the general capabilities described here.

What does a professional DTP scanner look like?

Unfortunately the answer to this question is irrelevant, since in our opinion professional DTP scanners are not yet on the market. If such a scanner were to exist, we would not have been compelled to write this book. This does not mean that you

cannot attain professional-quality pictures – quite the opposite. However, there are still many details, very important for everyday work, which have not been solved to an optimum degree. Therefore, we still have to wait for the professional scanner to appear.

Scanner and image processing software

At least as important as the scanner is the scanner software. It is beyond our scope here to discuss all the software available. However, we do wish to make a few basic remarks. The most important is that pictures may be scanned conveniently and accurately. You will perhaps laugh at this sentence now. Unfortunately, it is no laughing matter. Scanner software exists which possesses outstanding capabilities. Pictures may be retouched and you can draw, delete and add text. Everything is possible, with one exception: namely, reasonable scanning.

Using the so-called 'pre-scan' a complete page can be scanned at low resolution and displayed on the screen. The section of the picture which you wish to scan may now be enlarged with a frame precisely placed on it. During the actual scanning process it is just this section of the picture which is scanned very accurately. In order to evaluate the scan this section may then be further enlarged or reduced. However, scanner software exists which displays the pre-scan so small on the screen that it is hardly possible to define the section precisely. Often it is not even possible to enlarge or reduce it. If you wish to move the frame a little you may discover that even this is not possible. The complete frame must therefore be placed again and, of course, it is again not correct as the picture is too small to define the section precisely. In addition, setting the picture is difficult if the settings are not in single steps, but may only be adjusted roughly by bars. You then have to carry out many trials until the correct value is found. The software may also not remember the last value set, etc. Therefore, many irritations occur and often you only discover them after having purchased the scanner.

These examples are certainly only a small sample of many. They should, however, serve to emphasise that, when you are buying a scanner, do not let yourself be blinded by the multiple functions, but concentrate instead on the most important issues. Bear in mind that usually better special programs exist for picture retouching and similar functions. Make sure that you are shown the basic functions. In examining these remain critical and do not assume that the program will be able to do everything that you expect in normal performance.

Carry out a pre-scan, enlarge the picture section, set a frame, move the frame a little and then scan at a higher accuracy. Afterwards, enlarge and reduce the scan and adjust the picture settings for greyscale pictures. It should be possible to change the following picture settings: minimum and maximum brightness, gradation and inverse

video representation. The most important function here is the setting of the gradation. This should be capable of being set interactively via inputs and outputs (at least five support points). If this is not the case, the gradation may simply be considered as a novelty for the printout on a laser printer. For example, good retouching programs are available, but one of their shortcomings is that the picture settings cannot be reproduced. The most important aspect of image processing is taking the tone increase into account, which is produced when printing. If such a correction is carried out, good-quality pictures may be attained even without manual image processing.

Unfortunately, programs also exist where the brightness cannot be entered normally as a percentage, but only as a value in steps from 0 to 255. We consider these typical computer means of data entry completely superfluous, annoying and confusing. Specifications such as pels (number of pixels) are also not user-friendly, since users are not interested in numbers of pixels, but only in dimensions expressed in centimetres and resolutions expressed in dpi. In any case, you gain the impression with some programs that more value is placed on demonstration effects than on everyday use. For example, the effects of drawing gradation curves freehand may be demonstrated superbly. However, if the facility for storing and changing the gradation with the aid of support points is lacking, then such a program can hardly be recommended, unless it offers additional facilities which other programs do not possess.

A further important feature is the capability of interpolating data and subsequently changing the resolution. This is required for line drawings as well as for greyscale pictures. Associated with this feature are filter functions (smoothing, sharpening) which should also be available. It is important too that real grey levels may be represented on the screen with 64 levels at a resolution of at least 640 × 400 dots. Associated with this is the requirement that the screen should be capable of calibration. In image processing all changes (brightness, contrast and gradation) should be displayed immediately on the screen whilst being set (with the mouse). Programs where 'O.K.' must be clicked upon each time following a picture setting, and perhaps before the screen contents are even updated, are unsuitable for working smoothly and effortlessly. In the greyscale representation of a picture the progress of a curve should be blended in the diagram. Besides these basic functions, an image processing program should possess a comprehensive range of retouching facilities. Therefore, depending on the program, parts of or whole pictures may be extracted, changed, distorted or filtered.

We have described several important basic functions, which are not yet available in one unified program. Therefore, you will not be able to avoid completely specifying your requirements and evaluating which program best matches them. As you have certainly found out already, we rarely recommend particular programs in this book, even though we naturally have our own preferences. The problem here is that with a new version of a program the evaluation may change and we would have then

described an out-of-date program. We therefore consider it to be more sensible to make you aware of a few important characteristics, so that you can carry out your own assessment. Finally, we would just like to point out that anyone who buys an expensive scanner, without having a scanned (using this scanner), photoset and printed result presented beforehand, is really to blame if problems occur later.

As a 'pie in the sky' wish we imagine having a scanner software product orientated solely for the user. No longer would the resolution have to be entered, just the quality of the desired result. The scanner would have to do everything else itself automatically.

DTP software

A wide range of software exists for DTP applications. However, what actually is DTP software? Depending on the requirements several types of software packages belong to this category. The boundaries of DTP are also becoming blurred. Powerful word processing programs contain DTP elements and, conversely, DTP software contains word processing facilities. As mentioned already at the outset, DTP software takes texts and pictures from different programs and enables them to be integrated. Using DTP software and suitable hardware the layout may be viewed at almost actual size on the screen. Texts and pictures are shown virtually as they will be printed later. So, you require not only the actual DTP software, but also programs for the creation of text and pictures. The individual programs used depend on the requirements specification.

A demanding basic package may, for example, comprise the following:

- DTP program
- Word processing program
- Basic presentation program
- Scanner software
- Vector-orientated drawing program for text manipulation, creating drawings and for vectorising scanned drawings
- Program for processing line drawings

There are virtually no limits on making a system subsequently more and more complete. For example, an interesting issue would be the further processing of video images. It is also recommended that you think carefully about the use of various fonts. The standard fonts offered by printers are certainly sufficient at the beginning.

However, a few new fonts should be added for the different types of applications in order, for example, to lay out titles in different ways or simply to have a wider selection for blocked text. In addition, you can add libraries for the most diverse reasons.

Once all the parts have been assembled, you may start and the results will be wonderful... Unfortunately, not quite, since you will not become an expert overnight. On one hand, you should have selected the correct devices and programs and, on the other hand, the benefits of expert support should not be underestimated. Since it cannot be kept a secret, we should say that it is not possible to proceed completely without teething problems. Good training from the start and continuous support can prevent many problems, and practice makes perfect!

Ergonomic viewpoints on the workstation

In our opinion the issue of ergonomics at the workstation also belongs to the discipline of DTP. Work study experts have discovered that about 40% of performance is lost through an unsuitable workplace. Therefore, it is worth while to give some guidance here. We do not wish to describe working organisations here so much as the strain on health due to unsuitable conditions at the workplace.

Anyone who gets headaches or red and watering eyes from the screen is no longer capable of producing good work. However, it is not just technical aids which are required here, but also measures you can take yourself. If you sit for 8 hours a day (2000 hours a year) at the workplace, minor improvements can often bring a major benefit. Backaches and headaches should not be inevitable at the workstation.

The cause of headaches and tiredness is often continuous noise and demanding too much from the eyes. Eye strain may be caused by

- Incorrect screen and graphics card
- Bad arrangement of screen
- Unsuitable arrangement of workstation
- Incorrect lighting

The screen

The most important link between man and machine is the screen and keyboard. However, often too little attention is paid to the correct selection of a screen. It is surprising how much influence a suitable screen has on the well-being of a person, without the person always being aware of it. For example, the so-called flickering of

a screen may be stressful to health. Owing to the flickering the eye muscles are continually and almost imperceptibly strained. Even if this does not always lead directly to headaches, it does result in the eyes becoming tired early.

In addition, a continual change in brightness is a strain on the eyes. If you are continually glancing from a dark monitor to a light environment or an original document, you should not be surprised if you have eye problems. The difference in the levels of brightness between the screen, keyboard, manuscript and other fields of view should be as small as possible. This avoids the eye having to adjust itself continually to different light characteristics. For this reason a monitor with black characters on a white background is an optimum solution, providing that it does not flicker. How much a monitor flickers depends on the refresh rate, specified in hertz (Hz).

For an ergonomic monitor with little flickering both the monitor and the associated graphics card should have a refresh rate of at least 70 Hz. Recently monitors and graphics cards have become available which attain refresh rates of more than 80 Hz.

In the Civil Service and in companies which have their own departments responsible for ergonomics, usually only such monitors may now be used. The eyes do not notice flickering so much if the monitor is not turned up so bright. You are able to see the flickering most clearly if you turn your head to the side and look at the monitor from the corners of your eyes. If someone promotes a monitor as 'ergonomic' and then specifies a refresh rate of about 60 Hz, this is very insolent. In large monitors (e.g. 19") this problem is increased due to the large reflective area, so complete and double-page monitors, as well as their associated graphics cards, should attain a refresh rate of 75 Hz. However, in the specification of refresh rate it is the performance of the graphics card and not the monitor which is important for practical work.

How is flickering created?

Why does a screen flicker? Individual characters are written to the screen one after another. By the time the last letter is written at the bottom right corner, the first letter at the top left corner has already become a little darker, i.e. the individual characters need to be continually refreshed. If the time interval between the creation of the first and the last letter is too large, flickering is produced. Flickering is hardly noticeable, but leads to eye strain.

The higher the refresh rate the more often the individual characters are refreshed on the screen and hence the less the screen flickers. If a monitor and its associated graphics card attain a refresh rate of 70 Hz, this means that the picture on the monitor is displayed freshly 70 times per second.

By using a 'trick', for certain monitors it is possible to ensure that the eye is almost unaffected despite a low refresh rate. A special layer is painted on the screen which remains illuminated longer, so that the flickering is reduced. This illumination is disturbing in graphics where objects move quickly. In addition, this light layer is very expensive for both colour and black and white monitors and hence is not used for these. If you are working with inverse video representation, then even the longer illumination time cannot suppress the flickering. Monochrome monitors with green or amber screens stay illuminated longer and so may be driven using standard Hercules graphics cards which have a refresh rate of only 50 Hz.

As both colour and black and white monitors do not have the special layer which stays illuminated longer, these monitors need to be driven using a higher refresh rate. It is primarily the quality of the graphics card, and not the quality of the monitor, which is important here. The best monitor is of no use if it is controlled by a graphics card with a low refresh rate or if the software, for example, is not able to use fully the capabilities of the graphics card.

Graphics cards often attain either a high resolution or a high refresh rate, but usually not both at the same time.

High values may exist for the resolution and refresh rate in the data sheet, but in practical operation these values are not attained.

Let us have a closer look at a standard graphics card. EGA graphics cards only attain a refresh rate of 60 Hz. Even VGA graphics cards only attain a refresh rate of 70 Hz in text mode, but only 60 Hz in graphics mode. In order to attain 70 Hz in graphics mode with a VGA graphics card it needs to be operated in EGA mode. However, special graphics cards have recently appeared which behave as standard Hercules or standard VGA cards, but which operate even in graphics mode using a refresh rate of more than 70 Hz and so have very little flickering.

In addition, it is important for the screen to be highly anti-dazzle. The screen should also be located so that reflections are avoided. We do not wish to mention the screen resolution in detail here, as hopefully no one is working any longer with a CGA graphics card. In large monitors it is often possible to choose via suitable programs (screen drivers) whether work is to be carried out at a higher resolution or at a higher refresh rate.

Below is some technical data which monitors must attain in order to permit ergonomic working at 70 or 72 Hz. Attention must be paid to the fact that some multisync monitors may only be set to several fixed frequencies and that a refresh rate of 70 Hz is only realised for text mode and not graphics mode. The values listed are those with which the graphics card and monitor must comply:

Resolution	Refresh rate	Line frequency	Video bandwidth	Ergonomic
640×480	70 Hz	35.0 kHz	26 MHz	yes
640×480	72 Hz	38.4 kHz	27 MHz	yes
800×600	70 Hz	45.0 kHz	1 MHz	yes
800×600	72 Hz	48.0 kHz	42 MHz	yes
1024×768	70 Hz	56.0 kHz	66 MHz	yes
1024×768	72 Hz	58,0 kHz	68 MHz	yes
1024×768	76 Hz	62.0 kHz	72 MHz	yes
640×480	60 Hz	31.5 kHz	22 MHz	no
800×600	60 Hz	38.0 kHz	35 MHz	no
1024×768	60 Hz	49.0 kHz	57 MHz	no

Poor arrangement of the screen

The top of the screen should not be higher than eye level when you are sat upright. If a screen is positioned too high it causes you to hold your head unnaturally. The cause of this may, for example, be a screen support arm which is set too high or the screen being positioned on top of the computer module. Simply because of the noise generated and the space required the computer should not be placed on the desk. By using extension leads and a special stand the computer may be positioned vertically on the floor. In addition, it is sensible not to work in front of a window. However, windows behind should also be avoided because of reflections in the screen. Venetian blinds are often a relatively good solution to attain the correct light characteristics. If, for example, you need to type text whilst copying an original, you should use a stand which is set up directly next to the screen. Essentially, all movements (even eye movements) which you need to carry out often should be as short as possible. For this reason the screen should not be arranged too high so that it is in the same field of view as the keyboard.

Workplace lighting

It must not be forgotten that it is not only the monitor, but also the workplace lighting which can flicker. Any information which strains the eye will only be processed properly with difficulty. Disturbing reflections, too weak or too bright a light source are, besides noise, the worst enemy to working. In addition, flickering may emanate from fluorescent tubes. Furthermore, issues are not only related to the correct brightness for the room: even the colour of the light and the colour temperature should be right. Compare the light from a fluorescent light with a normal bulb. As a temporary measure better light characteristics may often be attained using an additional desk lamp (light bulb).

Sitting dynamics

It may irritate some of you to know that we are even going to discuss backache. However, anyone who carries out intensive screen activities over a long period will perhaps be grateful for a few tips. The prerequisite for ergonomically correct working is associated with the correct layout of the complete office. Besides the correct working devices and light characteristics, ergonomic office furniture also naturally helps to maintain your health.

The most important requirement today is that of sitting dynamics: an expression for the interplay between sitting surface and seat back which ensures that a permanent support is provided for the spine and back. The back of the chair should follow the movement of the sitting person by means of springs. This supports the body even when a bent-forward working posture is adopted. A special shape and design of the upholstery enhances this effect. When sitting, the pressure on the spinal discs is twice as severe as when standing and eight times as severe as when lying down. Backache, disc damage and spinal damage due to bending over are the results. It is, however, a mistake to believe that it is only through the possession of a good seat that these stresses are completely removed. The correct furniture is a help as well as the first important step in maintaining health.

An ergonomically correctly designed seat should not be missing from any workplace.
Photograph: Martin Stoll GmbH

It cannot be denied. Anyone who sits for 8 hours a day suffers severely from the lack of movement. Stress not only activates the brain, but stimulates the entire body. If the body is not able to stimulate relief then this leads to damage to the heart, circulation, vascular dilation, muscles and limbs. Movement, i.e. a change in the working posture between sitting and standing, increases the strain on the whole body and prevents aches. Good gymnastic exercises also help to strengthen the muscles in the stomach and back areas. In addition you should try to improve your posture during the day and always sit upright.

Index

A

Advertisements 271
Advertising 270
ANSI 282
Archiving 49
Area covering 137
ASCII table 281
Auto-tracing 54

B

Black and white mode 51
Bright light 155
Brightness 155, 157
    Limit of Brightness 34

C

Calibration
    Monitor . . . . . 161 - 170
    Photosetter 101
    Scanner 159
Camera-ready copy 95, 97
    Special paper 97
    Printing film 97
Camera scanner 44, 301
CCD sensors 34, 38, 42
Cicero point system 272
Circular dot . . . . . 186 - 188
Colour copying 119
Colour density 117
Colour layer 117
Colour pictures 240

Colour scanner . . . 41
Colour separation . . . 118
Colours . . . 116 - 117
Contrast . . . 155, 157
Curvature . . . 156

D

Data reduction . . . 89 - 90, 197
Data storage . . . 296
Data transfer . . . 296
Densitometer . . . 123, 139, 141
Density . . . 161
Density: conversion table . . . 263
Descender . . . 272
Didot point system . . . 272
Digital printing machines . . . 122
Digitiser . . . 233
Direct printing film . . . 134
Distortions . . . 40
Dithering . . . 35, 76
DMV Laser printer . . . 97, 121, 298
Dot shape . . . 186
Drop-out colours . . . 41, 163
DTP Software . . . 304

E

Effective scanning resolution . . . 54
Elliptical dot . . . 186, 188

F

Factor, Scanning and image resolution . . . 81
FBAS . . . 242
Feeder scanner . . . 44
File sending . . . 99
Film . . . 101
  Hybrid film . . . 101
  Line film . . . 101
  Lith film . . . 101
Flat-bed scanner . . . 299

Flickering . . . 292
Flourescent light . . . 41
Fonts . . . 272
Formulae
  Conversion of inches into centimetres . . . 254
  Effective scanning resolution . . . 54, 84
  Greyscale pictures . . . 254
  Image resolution . . . 83
  Image size . . . 54, 83
  Image Size (video) . . . 236
  Number of grey levels . . . 72
  Raster width (video) . . . 236
  Relative printing contrast . . . 143
  Scanning resolution for greyscale pictures . . . 70
  Scanning resolution for line drawings . . . 53
Four-colour printing . . . 119
Frequency of grid raster . . . 61
Frequency of half-tone raster . . . 61
Full-tone colour . . . 117
Full-tone density . . . 123, 139

G

Gamma curve . . . 156
Gradation . . . 156, 165, 167, 228
Graphic card . . . 293 - 294
Grey levels . . . 36, 71, 243
  Grey level scanner . . . 34, 35
  Grey level simulation . . . 59
  Grey levels, monitor . . . 153, 293
  Grey levels, printing . . . 36
Greyscale wedge . . . 160 - 161
Grid raster . . . 61

H

Half-tone increase . . . 102
Half-tone raster . . . 61
HDTV . . . 242
High resolution laser printer . . . 97, 121, 298
Hosiden signal . . . 242
Hybrid film . . . 101

I

Image size . . . 53
Image and scanning resolution . . . 80, 82
Image processing . . . 171, 195
Image processing software . . . 302
Image size, video . . . 235
Imposition . . . 146, 271
Inch/centimetre conversion . . . 61
Intensity . . . 155
Interpolation . . . 89, 91, 175

K

Kerning . . . 279

L

L/cm (Lines per centimetre) . . . 61
Laser printer . . . 69, 121, 297
  DMV Technology . . . 121, 298
  RET . . . 298
Layout . . . 267
Leading . . . 277
Letter spacing table . . . 99
Letter widths . . . 277
Light . . . 116
Light and shade . . . 177, 198
Light source . . . 40
Light spectrum . . . 41
Light trap . . . 137
Lighting . . . 308
Limit of Brightness . . . 51
Limit-check . . . 55
Line and page breaks . . . 99
Line drawings . . . 51
Line film . . . 101
Line spacing (leading) . . . 277
Lines . . . 100
Lith film . . . 101
Lpi (Lines per inch) . . . 61

M

Manipulations 228
Mean value formation 93
Medium tone 155
Memory capacity, greyscale pictures 37
Memory requirement 84
Missing colours 41
Moiré 35, 85, 89
Monitor . . . . . 293 - 294
  Grey levels . . . . . 153 - 154, 293
  Refresh rate 292
  Resolution 294
  Size 294
  VGA 154
  Calibration . . . . . 161 -170

N

NTSC 242

O

Optical Character Recognition (OCR) 47
  OCR, Failure rate 43
Offset printing 131
Optimisation factor 81
Orphans 280
Output resolution . . . . . 36, 39, 45 - 46

P

PAL 234, 241
Paper 152
  Coated Paper 248
  Paper class 168
  Paper quality 135
  Summary 248
Peak light 155
Pels 303
Photosetting 101
  Photosetter 105

Photosetters, raster widths . . . 107
Cost . . . 103
Film . . . 101
Order . . . 103, 262
Photosetting studio . . . 101 - 104
Pica point system . . . 272
Pixel, size . . . 64
Pixel and vectors . . . 55
Pixel picture . . . 54
Plate camera . . . 132
Polarisation filter . . . 124
PostScript . . . 110
PostScript character set . . . 281
PostScript laser printer . . . 297
PostScript printing data . . . 98
Pre-scan . . . 43, 302
Printing
Computer to plate . . . 134
Digital printing machines . . . 122
Grey levels . . . 71
Offset . . . 131
Printing order . . . 255
Quality . . . 114, 139, 142
Quality control . . . 135, 141
Standardisation . . . 137
Video images . . . 237
Printing contrast . . . 143
Printing file . . . 100
Printing film . . . 133
Printing medium . . . 145 - 146, 148
Printing order . . . 144, 255
Printing plate . . . 133
Proof . . . 142

Q

Quads . . . 280

R

Raster . . . 61, 68
Special-effect raster . . . 191
Raster angle . . . 66

Raster counter . . . 79
Raster dots . . . 66
Raster Image Processor (RIP) . . . 105
Raster tone . . . 117, 137
Raster width . . . 61, 125 - 126, 249
Photosetter . . . 107, 110
Calculation . . . 253
Dimensions . . . 61
Preset raster widths . . . 78 - 79
Raster/vector conversion . . . 54
Rasterising . . . 35, 36, 76 - 77
Reading . . . 274
Refresh rate . . . 292 - 294, 306, 308
Resolution
Line drawings . . . 51
Grey levels . . . 35
Greyscale . . . 37, 39
Monitor . . . 154, 293 - 294, 308
Printer . . . 69
Scanner . . . 69
Scanning . . . 53, 176
Video . . . 233
Video, TV . . . 234
RET . . . 298
Retouching . . . 228
RGB . . . 118
RIP (Raster Image Processor) . . . 106

S

S-VHS . . . 235, 242
Scanner . . . 298
Camera scanner . . . 44, 301
Colour scanner . . . 41
Feeder scanner . . . 44
Flat-bed scanner . . . 44, 299
Lenses . . . 40
Light source . . . 41
Sources of Light . . . 40
Transparency adapter . . . 42
Scanner Types . . . 44
Scanner calibration . . . 159, 161
Scanner resolution . . . 36 - 38, 45, 88
Scanner software . . . 43, 302

Scanning
  Greyscale pictures . . . 243, 246
  Line drawings . . . 53
  Rasterized originals . . . 88
  Scanning and image resolution . . . 80
  Scanning negatives . . . 44
  Scanning resolution . . . 53, 176
  Scanning drawings . . . 51
  Scanning slides . . . 44
  Scanning velocity . . . 43
Screen arrangement . . . 308
SECAM . . . 242
Serifs . . . 275
Set print film . . . 134
Shade . . . 155
Sharpening . . . 179, 198, 200
Size of pixel . . . 64
SmartSizing . . . 89, 91, 175
Smoothing . . . 90, 185, 197
Spaces . . . 280
Special-effect raster . . . 191
Spot colour . . . 118
Square dot . . . 186, 189 - 190
Standardisation . . . 137, 172
Still video camera . . . 233

T

Text height . . . 272
TIFF . . . 77
Tone comparison . . . 178
Tone correction . . . 157, 159
  Correction of tone increase, tables . . . 257-261
Tone increase . . . 135, 138, 140, 156, 199, 255 - 258
Tone jump . . . 186 - 187
Transparency adapter . . . 42
Typographical point . . . 272

V

Vectors . . . 54
VGA Monitor . . . 154, 294
VHS . . . 235, 242

Video 235, 236
Video standards 241

W

Widows and orphans 280
Word spacing 277

Y

YCM 118